This book is due for return not later than the
last date stamped below, unless recalled sooner.

Sampling Theory in Fourier and Signal Analysis

Foundations

Sampling Theory in Fourier and Signal Analysis

Foundations

J. R. HIGGINS

Division of Mathematics and Statistics
Anglia Polytechnic University, Cambridge

CLARENDON PRESS · OXFORD

1996

Oxford University Press, Walton Street, Oxford OX2 6DP

Oxford New York
Athens Auckland Bangkok Bombay
Calcutta Cape Town Dar es Salaam Delhi
Florence Hong Kong Istanbul Karachi
Kuala Lumpur Madras Madrid Melbourne
Mexico City Nairobi Paris Singapore
Taipei Tokyo Toronto
and associated companies in
Berlin Ibadan

Oxford is a trade mark of Oxford University Press

Published in the United States
by Oxford University Press Inc., New York

A catalogue record for this book is available from the British Library

Library of Congress Cataloging in Publication Data
(Data available)

ISBN 0 19 859699 5

Typeset by the author using L^{A}T$_E$X

Printed in Great Britain by
Bookcraft (Bath) Ltd
Midsomer Norton, Avon

To my dear family, without whom ...

PREFACE

We study sampling theory because of its intrinsic mathematical interest and because of its importance for applications in the sciences and engineering. One feature stands out in both these activities as being of prime importance: for members of certain function classes there is an equivalence between the function (a continuous information source) on the one hand, and a suitable collection of its samples (a discrete information source) on the other.

One of the chief ways of effecting this equivalence mathematically is to expand the function into a series containing the samples. The main purpose of this book is to provide a broad-based introduction to the mathematical study of sampling series.

From the point of view of applications, the relationship between the samples (regarded as measurements of a real signal) and the function (regarded as the mathematical model of the signal) needs to be thought through in a more refined way. This interesting problem of signal modelling is taken up in Chapter 17.

Sampling theory developed rather sporadically up to the middle of the present century, but since then its development has been rapid and continuous. Today there are many variations on this basic sampling theme, the field is large and contains many sub-disciplines. It would be impossible to cover this large field in one volume, and so an exhaustive coverage of the subject is not intended. However, an attempt has been made to write a reasonably comprehensive and balanced account containing a fairly wide variety of topics.

Textual material has been gathered from a diverse range of sources, and much of it has not appeared in book form before.

A companion volume to the present one, containing several topics at a more advanced level, will appear in due course under the title "Sampling theory in Fourier and signal analysis: advanced topics". It will contain contributions from several specialists in the area, and will be edited by J.R. Higgins and R.L. Stens; it will be referred to simply as H–S throughout the present volume.

A general introduction to the subject will be found in Chapter 1. Then some prerequisite material on Fourier analysis and Hilbert space theory is collected in Chapters 2 and 3; almost no proofs are given here, but references to standard sources are included to help the reader who wishes to find more information. Thereafter, the general plan of the book is to treat finite sampling as the first main topic in Chapter 4, make the transition to the case of infinite sampling in Chapter 5, and then to introduce infinite sampling schemes in the context of Bernstein and Paley–Wiener classes in Chapter 6. Chapter 7 contains an extended coverage of Paley–Wiener spaces, which, together with Bernstein spaces constitute by far the most important types of function class that we encounter in the earlier stages of sampling theory. The book continues with chapters treat-

ing most of the standard topics in the theory. However, a few topics which
might claim the right to inclusion have in fact been omitted, largely because one
can hardly do them justice in one chapter. These include stochastic sampling
(planned for inclusion in H–S), and the connections between sampling theory
and numerical analysis (which has already been given an excellent treatment
in the book of Stenger (1993)). Again, although error theories associated with
sampling have been introduced in Chapter 11, our treatment hardly begins to
do justice to this large specialist topic.

The book ends, as we have mentioned, with a chapter on signal modelling;
this leads up to Landau's theorem on minimal sampling rates, one of the high
points of the theory.

Some useful Fourier and Hilbert transforms are collected in Appendices A
and B.

It is worth pointing out that Chapter 5 is devoted entirely to a theorem of
Hinsen and Klösters on the transition from finite to infinite sampling. Readers
will no doubt forgive the presence of a rather long and technical proof at an early
stage of the book; its omission would, in the author's opinion, leave a general
account of sampling theory seriously incomplete. Of course it could be omitted
at a first reading, but at a second it is definitely not to be missed.

From time to time it will be necessary to "import" certain results to support the text, since there is not room here to develop everything we shall need;
references are of course supplied with these imports.

The book is aimed towards a mathematical audience; graduate students,
and those of the scientific community who wish to inform themselves about the
basics of the theory, particularly those in mathematics, applied mathematics,
mathematical physics or communications engineering. Readers are assumed to
have a general background in real and complex analysis, and some functional
analysis.

The book would readily lend itself to forming the basis for a one semester
graduate course. Students could be assigned chapters 1, 2 and 3 for self-study,
and then lectures could begin with Chapter 4 and continue with Chapters 6
and 10. Those classes wanting a more theoretical approach could continue with
Chapters 8, 9, 14 and 15. Those wanting a more applied flavour could do likewise
with Chapters 11, 12, 13 and 14. Both kinds of class could finish with Chapters
16 and 17.

Material not already covered in the first semester could form the basis of a
second semester course of either type, supplemented as required with topics from
H–S.

ACKNOWLEDGEMENTS

I would like to thank my wife Nan for her support of this endeavour, and in particular for her help in preparing the index and the bibliography.

Grateful thanks are extended to the following people, who kindly read parts of the manuscript and made valuable criticisms: Michael Beaty, Paul Butzer, Lorne Campbell, Neville Dean, Maurice Dodson, Adelheid Fischer, Gabriele Nasri-Roudsari (*geb.* Schöttler), Tibor Pogáni, Rolf Stens, Wayne Walker and Marcel Zwaan.

It is a pleasure also to thank Michael Beaty and Maurice Dodson for many fruitful conversations about sampling theory.

In addition, particular thanks go to Michael Beaty for his help in setting up a LaTeX system of document preparation, and for his advice on its use. This volume was typed by its author in *AMS* LaTeX.

Boz Kempski's support is also gratefully acknowledged.

Finally, special thanks go to Paul Butzer and to Maurice Dodson for their friendship and their continued help and encouragement, all of which has been of immense value to me over the years. It is doubtful if this book would have come into existence without them.

CONTENTS

1

AN INTRODUCTION TO SAMPLING THEORY

Let us suppose that a function f is defined for every point of some domain D, and has a series representation there of the form

$$f(t) = \sum_n f(\lambda_n) S_n(t), \qquad (1.0.1)$$

in which $\{\lambda_n\}$ is a collection of points of D, and $\{S_n\}$ is some set of suitable expansion functions. Such an expansion is called a *sampling series*, and the first thing we notice about it, the property that gives it its name, is that f is represented in its entirety in terms of its values, that is, its *samples*, at just a discrete subset of its domain. Series of this kind and their generalizations and extensions form the subject matter of this book.

On the theoretical side, sampling theory methods interact with fields such as interpolation and approximation, special functions and the expansion of special functions in series, expansions associated with eigenvalue problems, constructive function theory, entire functions and the Paley–Wiener theory, numerical analysis and harmonic analysis.

Furthermore, an important special kind of sampling series, the *cardinal* series which is introduced below, is known to be equivalent to other fundamental formulae of analysis in appropriate formulations, including the Cauchy integral formula and Poisson's summation formula. Another grouping of equivalent formulae, in appropriate formulations, consists of the cardinal series, the summation formulae of Poisson, Abel–Plana and Euler–MacLaurin, and the functional equation for the Riemann zeta-function (this topic will be discussed in §9.2). P.L. Butzer and his colleagues, to whom much of this work is due, are fond of quoting Salomon Bochner, who conjectured in 1978 that "the Poisson summation formula and Cauchy's integral and residue formulas are two different aspects of a comprehensive broad gauged duality formula which lies athwart most of analysis."

On the more applied side, sampling theory has seen application in such fields as systems theory, prediction theory, information theory, stochastic processes, optics, spectroscopy and two-dimensional image processing, and there are important connections in the relatively new development of multi-resolution analysis and wavelets.

In recent years iterative methods for the practical implementation of sampling theory have been developed; a comprehensive account can be found in Feichtinger and Gröchenig (1992).

A major application of sampling theory is in signal analysis. Here it provides the theoretical basis for modern pulse code modulation communication systems, having been introduced by Kotel'nikov in 1933 and Shannon in 1949. Its usefulness stems from the fact that in many situations the information contained in a continuously varying signal is the same as that contained in its samples alone, or put a slightly different way, certain functions whose frequency content is bounded are equivalent to an information source with discrete time.

A greater appreciation of the broad scope of sampling theory can be obtained from, for example, the books of J.M. Whittaker (1935), Marks (1991, 1993), Bilinskis and Mikelsons (1992), Stenger (1993), Zayed (1993b), Hoskins and Sousa Pinto (1994) (particularly Chapter 3) and Walter (1994), and the survey articles of Jerri (1977), Lüke (1978), Butzer (1983), Higgins (1985), Vaaler (1985), Butzer *et al.* (1988), Dodson (1992), Marvasti (1993), Brown (1993) and Butzer and Nasri-Roudsari (to appear) and their voluminous bibliographies.

Sampling theory and its applications continue to provide active areas of research at the present time.

1.1 General introduction

One could hardly think of a simpler example of the sampling series (1.0.1) than to take a linear function f on $\mathbb{R}$, say $f(t) = at + b$, and two points, chosen for convenience to be $\lambda_0 = 0$ and $\lambda_1 = 1$. Then a moment's thought shows that $at + b = b(1 - t) + (a + b)t$, and that (1.0.1) becomes

$$f(t) = f(0)(1 - t) + f(1)t. \qquad (1.1.1)$$

Already present, even in this extremely simple example, are features that run through all our work on sampling series. One of them is the feature of universality. Every member of the proposed class, in this case polynomials of degree one, is representable in the series (1.1.1), the expansion functions $S_0(t) = 1 - t$ and $S_1(t) = t$ being independent of f. This kind of universality is very desirable in a sampling theorem.

Another feature is that there is present in our simple example a certain duality, a theme to which we shall return in a moment. On the one hand, a straight line determines any two of its points; on the other, any two points determine a straight line. This raises a practical consideration; the closer together the two points are, the less certain can we be of which straight line is being determined. It is indeed a good idea to keep sample points as well separated as possible (this will be formalized in Definition 10.1).

Let us expand a little on the notion of duality that we just noticed in our simple example. We really confront two closely related problems here. One of them is as follows. Suppose that we are given a certain class of functions which are defined on some common domain: can we find a discrete subset Λ of this domain such that every member of the class is determined uniquely by the collection of values that it takes on Λ? And if this is the case, how can we recover such a function completely using these 'samples' only? This is the problem of *sampling and reconstruction*.

The dual problem is a problem of *interpolation*. We are again given a class of functions defined on a certain domain, but this time a set of interpolation points belonging to the domain is prescribed, together with a value assigned at each point. We ask whether there is a member, perhaps unique, of the given class taking the assigned value at each point, and if so, how it can be constructed.

When only finitely many points are involved there is rather little difference between these two; but when we pass to the case of infinitely many points there is a substantial difference. This will certainly be clear by the time we reach Definition 10.2.

A historical review of finite sampling and interpolation is to be found in §1.2, and here the famous Lagrange interpolation formula is presented. The topic is further developed in Chapter 4, but it is when (1.0.1) is an infinite series that the problems hold most interest. After a completely rigorous discussion of the passage to the case of infinite sampling series in Chapter 5, the remainder of the book is concerned exclusively with the infinite case.

$$* \qquad * \qquad *$$

It is quite natural that sampling series should arise in a Fourier analysis environment. Let us consider the following heuristic argument (the necessary background in Fourier analysis will be found in Chapter 2).

Let g be periodic with period 2π and let it generate a formal Fourier series, so that we have

$$g(x) \sim \frac{1}{\sqrt{2\pi}} \sum_{n \in \mathbb{Z}} d_n e^{-ixn}.$$

Now let us multiply both members of this relation by $e^{ixt}/\sqrt{2\pi}$ and integrate over $[-\pi, \pi]$. The integral required on the right-hand side is

$$\frac{1}{2\pi} \int_{-\pi}^{\pi} e^{ix(t-n)} \, dx = \frac{\sin \pi(t-n)}{\pi(t-n)},$$

so the right-hand side becomes

$$\sum_{n \in \mathbb{Z}} d_n \frac{\sin \pi(t-n)}{\pi(t-n)}.$$

The left-hand side is

$$\frac{1}{\sqrt{2\pi}} \int_{-\pi}^{\pi} g(x) e^{ixt} \, dx. \qquad (1.1.2)$$

This function, which we could call $f(t)$, has the appearance of being the inverse Fourier transform of a function that coincides with g over $[-\pi, \pi]$ and vanishes elsewhere. We can therefore think of f as a function that contains low frequency terms up to a certain "cut-off" within its range of frequencies, and after that is devoid of all higher frequencies. By "range of frequencies" we shall understand the support of the Fourier transform of f; the Fourier transform will need an appropriate interpretation in context. Such a function is often called a *low-pass signal*. More generally we have:

Definition 1.1 *Any function whose range of frequencies is confined to a bounded set B (which may not be a single interval but could be a union of several components, even in higher dimensions) is called band-limited to B. We refer to B by the name band-region.*

We shall adopt a useful piece of notation.
Definition 1.2

$$\operatorname{sinc} v := \begin{cases} \dfrac{\sin \pi v}{\pi v}, & v \neq 0; \\ 1, & v = 0. \end{cases}$$

The name "sinc"[1] is common in engineering literature, and we shall make much use of it from now on.

The result of our heuristic deliberations up to this point can be summarized as follows. If

$$f(t) = \frac{1}{\sqrt{2\pi}} \int_{-\pi}^{\pi} g(x)e^{ixt}\,dx, \tag{1.1.3}$$

then, still in the sense of formal equality,

$$f(t) = \sum_{n \in \mathbb{Z}} d_n \frac{\sin \pi(t-n)}{\pi(t-n)} = \sum_{n \in \mathbb{Z}} d_n \operatorname{sinc}(t-n). \tag{1.1.4}$$

Here, $\{d_n\}$ are the Fourier coefficients of g (§2.1); but these are obtained (as 2.1.3 would suggest) by placing $t = n$ on the right-hand side in (1.1.3), and from the left-hand side this means that $d_n = f(n)$.

There is another way of recognizing the coefficients as samples of f. From Definition 1.2 it follows that whenever $m, n \in \mathbb{Z}$, $\operatorname{sinc}(m-n) = \delta_{mn}$[2] (this is often called the *interpolatory property* of the functions $\operatorname{sinc}(t-n)$), so that if (1.1.4) converges pointwise, then when we put $t = m$ every term of the series vanishes except the mth, which is $f(m)$.

We can now write our sampling series (1.1.4) in the form

$$f(t) = \sum_{n \in \mathbb{Z}} f(n) \operatorname{sinc}(t-n). \tag{1.1.5}$$

---

[1] The name is usually held to be short for the Latin *sinus cardinalis*. It was introduced by Woodward (1953, p. 29), although it is not certain whether this is the earliest occurrence.
[2] Kronecker's delta symbol: δ_{mn} equals 1 if $m = n$, and equals 0 if $m \neq n$.

We refer generally to a series of the form (1.1.4) as a *cardinal series*. Whenever the series (1.1.5) converges to f we shall say that $f(t)$ is represented by its cardinal series. This series is clearly a special case of (1.0.1) in which the sample points are taken to be equally spaced. More generally, we shall refer to the scaled version

$$f(t) = \sum_{n \in \mathbb{Z}} f\left(\frac{n}{w}\right) \operatorname{sinc}(wt - n), \qquad (1.1.6)$$

in which $w > 0$, as a cardinal series.

Clearly, if (1.1.5) is unconditionally convergent it can be written in the equivalent form

$$f(t) = \frac{\sin \pi t}{\pi} \left\{ \frac{f(0)}{t} + \sum_{n=1}^{\infty} (-1)^n \left(\frac{f(n)}{t-n} + \frac{f(-n)}{t+n} \right) \right\}. \qquad (1.1.7)$$

When f is an even function (1.1.7) reduces to the even form

$$f(t) = \frac{\sin \pi t}{\pi} \left\{ \frac{f(0)}{t} + 2t \sum_{n=1}^{\infty} \frac{(-1)^n f(n)}{t^2 - n^2} \right\}, \qquad (1.1.8)$$

and when f is odd it reduces to the odd form

$$f(t) = \frac{2 \sin \pi t}{\pi} \sum_{n=1}^{\infty} \frac{(-1)^n n f(n)}{t^2 - n^2}. \qquad (1.1.9)$$

Many other names for the cardinal series can be found in the literature. It has been called Whittaker's cardinal series, the Shannon sampling theorem, the Whittaker–Shannon sampling series, or theorem; and when it became known in the western world that the Russian electrical engineer Kotel'nikov had discovered it independently in 1933, his name began to be added to the title as well. To complicate matters further, there are even more people who have legitimate claims to be included, as we shall see in the short historical survey in §1.2. Rather than try to find a title that does justice to all claimants, we shall simply refer to a series of the form (1.0.1) and its variants as a *sampling series*, and continue to refer to the special case (1.1.6) (and the less precise form in Theorem 1.3 below) as a *cardinal series*,[3] a name that can at least claim historical precedence over the others.

$$* \quad * \quad *$$

The sense in which pointwise convergence of a sampling series is to be understood must now be established. Our assumption will be that convergence is

---

[3] The name begins to appear in the works of J.M. Whittaker in the 1920s.

taken in the sense of the limit of symmetric partial sums, that is, in the sense that

$$\lim_{N \to \infty} \sum_{|n| \leq N} f(\lambda_n) S_n(t)$$

exists for all suitable t.

Actually, we shall discover that it is usual for a cardinal series to converge absolutely, and then of course the sense of convergence is immaterial. A criterion for absolute convergence follows in the next theorem. However, absolute convergence is not always guaranteed, as we shall find in §§6.5 and 6.8 for example, so we do need a working assumption. Again, the same assumption is usually adopted for trigonometric series in exponential form, and since we shall use these in some of our methods, consistency in this respect is essential.

When can we expect a cardinal series to converge absolutely? The following theorem gives a simple necessary and sufficient condition.

Theorem 1.3 (Absolute Convergence Principle) *The cardinal series*

$$\sum_{n \in \mathbb{Z}} a_n \frac{\sin \pi(wt - n)}{\pi(wt - n)}$$

converges absolutely, hence unconditionally, for every $t \in \mathbb{C}$ if and only if

$$\sum_{n \in \mathbb{Z}}{}' \left| \frac{a_n}{n} \right| < \infty. \tag{1.1.10}$$

(The prime is the standard notation indicating summation over non-zero values of n.)

Proof Since the series reduces to a single term if $t = m/w$, where m is any integer, we can assume throughout the proof that t is not of this kind.

Suppose that such a t is fixed in $\mathbb{R}$; then we can find a constant $D < 1$, and an integer M so large that $|wt/n| < D$ whenever $|n| \geq M$. Then

$$\left| \frac{a_n(-1)^n}{wt - n} \right| \leq \left| \frac{a_n}{n} \right| \frac{1}{1 - \left| \frac{wt}{n} \right|} < \left| \frac{a_n}{n} \right| \frac{1}{1 - D}, \qquad n \geq M.$$

Hence if $\sum' |a_n/n| < \infty$, the cardinal series converges absolutely for every real t.

Conversely, if the cardinal series converges absolutely, we have

$$\sum_{n \in \mathbb{Z}}{}' \left| \frac{a_n}{wt - n} \right| < \infty.$$

Now, for a fixed value of t,

$$\left|\frac{a_n}{wt - n}\right| \geq \frac{|a_n|}{|wt| + |n|} > \frac{|a_n|}{2|n|}$$

for all n such that $|n| > |wt|$; hence $\sum' |a_n/n| < \infty$.

For example, if $\sum |a_n|^2 < \infty$ then the Cauchy–Schwarz inequality for sums applied either to the series directly, or to (1.1.10), shows at once that the cardinal series is absolutely convergent.

1.2 Interpolation and sampling from the seventeenth century to the mid-twentieth century — a brief review

It has already been mentioned that the dual problems of reconstruction and of interpolation do not differ greatly when finitely many points only are involved. It was certainly this case which first arose historically, and so it will be of no little interest to go back to the early seventeenth century, even if only very briefly, to trace these matters back to their origins. The citations omitted in this section can all be found in Higgins (1985).

Interpolation arose because of the need to calculate values of a given function which were intermediate between tabulated values. Using values that had already been calculated would alleviate some of the drudgery of calculating the new value all over again from the beginning. Methods were developed for this by Briggs, using successive and modified differences; these were published in London in 1624 in his *Arithmetica Logarithmica* and seem to be the first contribution to the subject.

Suggestive ideas of Wallis had some influence in the 1650s, and he can be credited with introducing the word "interpolation"[4] but it was Gregory and Newton who first recognized the polynomial nature of the interpolant based on successive differences. Indeed, in a letter dated 23 November 1670 to Collins,[5] Gregory gave the first interpolation series, which took the form

$$f(0) + t\Delta f(0) + \frac{t(t-1)}{2!}\Delta^2 f(0) + \frac{t(t-1)(t-2)}{3!}\Delta^3 f(0) +$$
$$\cdots + \frac{t(t-1)\ldots(t-M+1)}{M!}\Delta^M f(0),$$

where

---

[4] In *Arithmetica Infinitorum*, Oxford, 1655.

[5] John Collins (1625–1683), largely self-taught as a mathematician, was the author of works on navigation and accountancy as well as mathematics. At the time of this correspondence with Gregory he was a prominent member of the newly founded Royal Society of London and corresponded with several leading mathematicians of his day, apparently acting as a kind of intermediary between them. He is described in the *Dictionary of National Biography*, Volume XI, 1887, as " ... constantly stimulating others to useful enquiries".

$$\Delta f(0) = f(1) - f(0);$$
$$\Delta^2 f(0) = \Delta f(1) - \Delta f(0) = f(2) - 2f(1) + f(0);$$
$$\cdots$$

This formula solves the finite interpolation problem of finding a polynomial of degree M, or less, which takes the given value $f(0)$ when $t = 0$, the value $f(1)$ when $t = 1$, etc., and the value $f(M)$ when $t = M$.

In the 1670s Newton introduced divided differences and adjusted divided differences, and subsumed all the interpolation formulae known at the time under a single rule. Both his and Gregory's methods " ... derive from taking the polynomial as a close approximation to the (continuous) function to be interpolated" (Whiteside 1961, p. 250).

Let us consider the identity

$$f(\lambda_0) + (t - \lambda_0)f(\lambda_0, \lambda_1) + \cdots + (t - \lambda_0)\ldots(t - \lambda_{M-1})f(\lambda_0, \ldots, \lambda_M)$$
$$\equiv \sum_{j=0}^{M} f(\lambda_j) \frac{G_M(t)}{G'_M(\lambda_j)(t - \lambda_j)}, \qquad (1.2.1)$$

in which

$$f(\lambda_0, \lambda_1) = \frac{f(\lambda_1) - f(\lambda_0)}{\lambda_1 - \lambda_0}$$
$$f(\lambda_0, \lambda_1, \lambda_2) = \frac{f(\lambda_1, \lambda_2) - f(\lambda_0, \lambda_1)}{\lambda_2 - \lambda_0},$$
$$\ldots,$$

and

$$G_M(t) = \prod_{j=0}^{M} \left(1 - \frac{t}{\lambda_j}\right).$$

The two sides of (1.2.1) are identical because both are polynomials of degree M taking the value $f(\lambda_0)$ when $t = \lambda_0$, $f(\lambda_1)$ when $t = \lambda_1$, etc., and $f(\lambda_M)$ when $t = \lambda_M$.

Here the interpolation points are no longer integers but consist of arbitrary real numbers $\{\lambda_j\}$, $j \in \mathbb{N}_0$. The left-hand side of this identity is known as Newton's divided difference formula, and clearly reduces to Gregory's formula when $\lambda_j = j$, $j \in \mathbb{N}_0$.

The right-hand side of (1.2.1) is known today as Lagrange's formula. It appears in a set of lectures on elementary mathematics that Lagrange gave at the Ecole Normale in Paris in 1795.[6] However, it had already been discovered by Waring (1779, p. 59), whose purpose was to construct the polynomial interpolant

---

[6]See, for example, the article by Netto and le Vavasseur in the *Encyclopédie des Sciences Mathématiques pures et appliqueés*, Tome I, volume 2, Gautier-Villars, Paris, 1907, p. 58.

" ... without having any recourse to finding the successive differences". This is a significant development because the Lagrange formula gives the polynomial interpolant directly in terms of the samples, and takes on the form of the sampling series (1.0.1).

*   *   *

Suppose now that a periodic interpolant is wanted rather than a polynomial one; or, in the spirit of the dualism mentioned above, suppose we wish to represent a periodic function of the form $p(t) = \sum_{j=-M}^{M} c_j e^{ijt}$ say, by a finite sampling series. An appropriate formula is

$$p(t) = \frac{1}{2M+1} \sum_{j=0}^{2M} p\left(\frac{2\pi j}{2M+1}\right) \frac{\sin\left[\left(\frac{2M+1}{2}\right)\left(t - \frac{2\pi j}{2M+1}\right)\right]}{\sin\frac{1}{2}\left(t - \frac{2\pi j}{2M+1}\right)}. \qquad (1.2.2)$$

This finite sampling series was given by Cauchy in 1841, although there are earlier forms in the work of Gauss, going back to about 1805.[7] A more general form can be found in Dodson and Silva (1985).

*   *   *

During the nineteenth century the passage from the case of finitely many points to the significantly more general case of infinitely many points $\{\lambda_n\}$ (with no finite point of accumulation) in Lagrange's formula had been made. This formula, for a function f taking the value a_n at λ_n, can be formally written in the form

$$f(z) = \sum_n a_n \frac{\phi(z)}{\phi'(\lambda_n)(z - \lambda_n)} \qquad (1.2.3)$$

where ϕ is now a function with simple zeros at $\{\lambda_n\}$. A more general formula is

$$\sum_n a_n \left(\frac{z}{\lambda_n}\right)^{s_n} \frac{\phi(z)}{\phi'(\lambda_n)(z - \lambda_n)},$$

where (s_n) is a sequence of integers chosen to ensure convergence (see, for example, J.M. Whittaker 1935, p. 3).

---

[7]See, for example, the article by Burkhardt in the *Encyclopédie des Sciences Mathématiques pures et appliqueés*, Tome II, volume 5, Gautier-Villars, Paris, 1907, p. 94.

By 1882 Cazzaniga was using such formulae for constructive purposes, with due regard to the convergence. In 1899 Borel took $\lambda_n = n$, and, using the fact that $\sin \pi z$ has the appropriate zeros, obtained the first known explicit statement of what we now call the cardinal series.

Another significant contribution, made by Borel in 1897, was to observe that if $f(t)$ is any function of the form (1.1.3), with a mild restriction on g, then the sample values $f(n)$, $n \in \mathbb{Z}$, determine f uniquely.

Taken together, these two contributions of Borel are really a solid first step towards sampling principles as we have come to know them in later times.

Before leaving the nineteenth century it should be pointed out that another, extended form of Lagrange interpolation at infinitely many points had also been developed, and had appeared in the work of Guichard by 1884. Special cases of this form will be found in Theorem 6.21(a) and Problem 6.9. It was taken up in more detail in the work of Valiron in 1925, where it is referred to as Lagrange interpolation with an associated "rank" number, a number that counts the additive terms preceding the summation. Today it is known as the interpolation series of Valiron.

$$* \quad * \quad *$$

During the first half of the twentieth century there was much emphasis in the literature on the cardinal series, a special case as we have seen of the Lagrange formula with the simplicity gained from taking equidistantly spaced points. The series was rediscovered in this form several times; for example, by Whipple in 1910 and independently by E.T. Whittaker in 1915, together with the band-limited nature of the sum. Butzer and Stens (1992) noticed that in 1920 Ogura was apparently the first to supply the cardinal series with a fully rigorous proof (a commentary on Ogura's work can be found in Zayed (1993, p. 19)). In the 1930s and later, several studies appeared in which the cardinal series and the Valiron series were used in such problems as the deduction of the behaviour of functions from their behaviour along a sequence of points.

In 1941 Hardy noticed that the cardinal series is an orthogonal expansion; this was the beginning of an important development in which sampling and reconstruction of functions is seen in terms of their expansions in various bases for Hilbert and Banach spaces. It will be an important theme throughout this book.

In 1949 Shannon introduced sampling methods into communication theory (he had in fact been anticipated by Kotel'nikov in 1933, but Kotel'nikov's work did not become widely known in the western world until after 1960 or thereabouts). It should also be mentioned that work of Raabe (1939) in this direction also predated that of Shannon. Applications in communication theory have proved to be very influential in the theory; they have stimulated, and indeed continue to stimulate, a vast amount of research.

Shannon's work was completed several years earlier than its publication date, but 1949 was a fruitful year and saw two other apparently independent suggestions introducing sampling methods to communication theory, from Weston and Someya.

Mention should also be made of a much earlier study of 1908 by de la Valleé Poussin (the subject of an interesting historical account by Butzer and Stens (to appear)), in which a rather different kind of sampling theory is used to treat functions that are time-limited rather than band-limited. A function of time is said to be *time-limited*, or *duration-limited*, if it vanishes outside a bounded interval, and here is a good place to mention that throughout this book independent variables will usually represent time and be denoted by t; the variable x will usually mean a frequency variable. We come to think in terms of a "time domain" and a "frequency domain".

We must not leave these brief historical remarks without mentioning that Beurling was apparently in possession of at least rudimentary sampling methods in the late 1930s, independently of other workers in the field; these included constructive methods involving not only samples of the object function but also samples of its derivative (derivative sampling will be treated in §9.3 and Problem 12.2). This remarkable fact was brought to light in an interesting survey by Vaaler (1985). Some of the topics treated there are planned for inclusion in H–S.

These historical remarks have only touched on some of the more significant events. Several historical sources are now available from which the interested reader can find more information; for example, the books and survey articles mentioned at the beginning of this chapter, and the other historical articles of Butzer *et al.* listed in the Bibliography.

1.3   Some further introductory remarks

This introductory chapter ends with a few further remarks about important features of sampling that will recur throughout our work.

First, we shall frequently, but by no means always, take our band-regions B to be symmetrically placed with respect to the origin. The reason is to be found in a property of the Fourier transform $f^\wedge$ of f (looking forward for a moment to the definitions of the transform in §2.2, where it is important to note that the unit of frequency is taken to be one radian per second). Now if f is real-valued, as it usually is for applications, we have

$$\overline{f^\wedge(x)} = \frac{1}{\sqrt{2\pi}} \int_{\mathbb{R}} f(t)e^{ixt}\, dt = f^\wedge(-x),$$

so that $|f^\wedge(x)|^2 = \overline{f^\wedge(x)}f^\wedge(x) = f^\wedge(-x)f^\wedge(x)$, an even function. Hence, if a real-valued f is band-limited, then its band-region B must be symmetric with respect to the origin. An example of a complex-valued function whose band-region is nevertheless symmetric can be found in Appendix A, no. 12.

Again, it is important to consider the *rate*, or *density*, at which sample points occur in the cardinal series. In the case of the "canonical" form of the series

(1.1.5), we see that this rate is one sample per second. More generally we have:

Definition 1.4 *The Nyquist sampling rate for a function band-limited to B is defined to be twice the least upper bound on the frequency spectrum of f, that is,* $2(\sup_{x \in B} x)/2\pi$, *the factor* 2π *appearing due to the units in which frequency is measured; namely, radians per second.*

When B is a single interval centered at the origin, the Nyquist sampling rate is evidently equal to $m(B)/2\pi$ (here and throughout the rest of this book $m(S)$ will denote the Lebesgue measure of the set S). Its importance was recognized in pioneering work of Nyquist in the later 1920s, and its significance, which will be brought out in the delta method below, stems from the fact that functions of the form (1.1.3) can be sampled and reconstructed at this rate or higher (over-sampling), and this is often done in practice, but stable sampling and reconstruction is not possible at a lower rate (under-sampling). A more complete discussion will be found in §10.1.

A question that is closely associated with these matters has been raised by Seip (1992, p. 322), who asks for " ... a general characterisation of the information needed to represent signals, ... ". While partial answers are known, this question is such a general one that a universal answer cannot be given at the present time. The topic is planned for inclusion in H–S.

Another point of general interest concerns the kind of functions that can be expected to have a representation in sampling series. In the treatment given in this book they will (generally speaking, and apart from the multi-dimensional case treated in Chapter 14) be entire functions of exponential type in the complex plane, satisfying an integrability condition or a polynomial growth condition along the real axis. By the Paley–Wiener theorem (§7.1), or one of its extensions, such functions will be band-limited in a suitable sense. Since a function of this kind is entire, it has the "rigid" property that the whole function is determined by its values on any subset of $\mathbb{C}$ containing an accumulation point. It cannot, therefore, vanish on a line segment without vanishing identically; in particular it cannot be time-limited (unless it is the null function and hence not very interesting). Again, it can be determined, for example, by analytic continuation from its values on any interval of $\mathbb{R}$ with positive length, no matter how small.

$$* \qquad * \qquad *$$

In a final remark we shall obtain our first direct derivation of the cardinal series representation for a band-limited function, employing the "delta method" . It is not fully rigorous but it is basic and natural, and has proved to be extremely fruitful; in fact, many sampling theorems owe their discovery to this remarkable method. From about 1950 onwards there was a tendency for sampling formulae to be derived using the delta method; only later was it felt desirable to find other, more rigorous, derivations. The method itself was put into a sound mathematical framework by Nashed and Walter (1991).

To sample a function f at equidistantly spaced points $\{n\tau\}$ we make a mathematical imitation of engineering practice, and subject it to a "train of deltas". Thus a sampled version f_s is formed as follows:

$$f_s(t) = f(t) \sum_{n \in \mathbb{Z}} \delta(t - n\tau) = \sum_{n \in \mathbb{Z}} f(n\tau)\delta(t - n\tau),$$

where τ is a parameter governing the sampling rate. This suggests that f itself can be represented as

$$f(t) = \sum_{n \in \mathbb{Z}} f(n\tau)g(t - n\tau), \tag{1.3.1}$$

in which g is a reconstruction function, independent of f. It is our task to find g and a suitable value for τ. First we note that (1.3.1) can be written

$$f(t) = \int_{\mathbb{R}} f(u)g(t - u) \sum_{n \in \mathbb{Z}} \delta(u - n\tau) \, du. \tag{1.3.2}$$

Next we quote a well-known Fourier series expansion for our periodic delta train:

$$\sum_{n \in \mathbb{Z}} \delta(u - n\tau) = \frac{1}{\tau} \sum_{n \in \mathbb{Z}} e^{-2\pi i n u/\tau}.$$

On rearranging (1.3.2) we find a summand consisting of $f(\cdot)e^{-2\pi i n \cdot/\tau}$ convolved with $g(\cdot)/\tau$. We take Fourier transforms on both sides of (1.3.2) (using the operational properties in Appendix A no. 3b and no. 7), obtaining

$$f^\wedge(x) = \frac{1}{\tau}\sqrt{2\pi}g^\wedge(x) \sum_{n \in \mathbb{Z}} f^\wedge\left(x + \frac{2n\pi}{\tau}\right). \tag{1.3.3}$$

On the right of this relationship we find repeated copies of the Fourier transform of f; this is called *spectrum repetition* in the engineering phraseology. However, there is only one copy on the left. Therefore to make sense of (1.3.3), $g^\wedge$ must be chosen so that it picks out just one copy of the spectrum from the repetitions on the right; of course, if these were to overlap our task would be impossible. Now suppose that f is band-limited to $[-\pi w, \pi w]$. Then we must give τ a value $1/w$ or greater to prevent the spectra overlapping; thus we can over-sample but we cannot under-sample, and when $\tau = 1/w$ we obtain the Nyquist sampling rate of Definition 1.4.

Now we can take $\sqrt{2\pi}g^\wedge(x)/\tau$ to be a "window" through which we can "see" that one copy of the spectrum corresponding to $n = 0$; so we must take

$$g^\wedge = \frac{1}{w\sqrt{2\pi}}\chi_{\pi w}(x),$$

where $\chi_{\pi w}$ denotes the characteristic function of the interval $[-\pi w, \pi w]$. Inverse Fourier transforms can now be taken, and using Appendix A, no. 8, we obtain

$$g(t) = \frac{\sin \pi w t}{\pi w t},$$

and (1.3.1) is the cardinal series.

The delta method makes spectrum repetition arise in a very natural way, and shows clearly how its optimal deployment, that is, its arrangement so that the whole frequency axis is covered leaving neither gaps nor overlaps, gives us the Nyquist sampling rate.

2

BACKGROUND IN FOURIER ANALYSIS

In this chapter we shall review those parts of classical Fourier analysis that will be needed. This material is collected here for convenience, and since it is extremely well known and available in any good book on the subject, no proofs will be given except in the section on Poisson's summation formula where a slightly more expanded discussion is given. The classical texts of Butzer and Nessel (1971), Titchmarsh (1948) and, for higher-dimensional Fourier analysis, Stein and Weiss (1971) are excellent source books. For an exposition of the fundamental ideas of Fourier analysis, and a broad range of applications, the book of Dym and McKean (1975) is highly recommended.

Throughout, the exponents p and q are related by $1/p + 1/q = 1$ and are said to be *conjugate* to each other.

Let X be a subset of $\mathbb{R}$ (from the topological point of view, X is taken to be locally compact).

Definition 2.1 *The symbol $L^p(X)$, $1 \leq p \leq \infty$, denotes the collection of all equivalence classes of functions whose pth power is Lebesgue integrable over X (essentially bounded over X in the case $p = \infty$), any two members of an equivalence class being equal almost everywhere.*

We have $L^r(X) \subset L^s(X)$, $s < r$, whenever $m(X) < \infty$.

The equivalence class containing the identically zero function is denoted by θ, and called the *null element* of $L^p(X)$. Often the notation $f \in L^p(X)$ will be used when f is a specific function rather than an equivalence class; the meaning will be clear in context.

Definition 2.2 *A norm on L^p is given by the real-valued function $f \mapsto \|f\|_p$, where*

$$\|f\|_p = \left\{ \int_X |f(t)|^p \, dt \right\}^{1/p}, \qquad 1 \leq p < \infty;$$
$$\|f\|_\infty = \operatorname*{ess\,sup}_{x \in \mathbb{R}} |f(x)|.$$

With this norm $L^p(X)$ is a Banach space. When $p = 2$ it is a Hilbert space with inner product

$$\langle f, g \rangle = \int_X f(t) \, \overline{g(t)} \, dt.$$

More details about Hilbert spaces are to be found in Chapter 3.

We shall need the very useful inequalities of Hölder and Minkowski. The basic results are as follows.

Theorem 2.3 (Hölder's inequality) *Let $1 \leq p \leq \infty$, and let $f \in L^p(\mathbb{R})$ and $g \in L^q(\mathbb{R})$, and suppose that neither f nor g is null. Then $fg \in L^1(\mathbb{R})$, and we have*

$$\|fg\|_1 \leq \|f\|_p \|g\|_q.$$

The case $p = q = 2$ is usually known as the Schwarz Inequality, or by some combination of the names Cauchy, Schwarz and Buniakovskii.

Theorem 2.4 (Minkowski's inequality) *Let $1 \leq p \leq \infty$, and let f and g both belong to $L^p(\mathbb{R})$. Then*

$$\|f + g\|_p \leq \|f\|_p + \|g\|_p.$$

Conditions giving equality in these two inequalities will not be needed; the interested reader can trace this information in, for example, Dunford and Schwartz (1988, p. 173).

2.1 The Fourier series

Let f be a complex-valued function on $\mathbb{R}$ which is periodic with period $2\pi w$. Let

$$c_\nu = \frac{1}{\sqrt{2\pi w}} \int_{-\pi w}^{\pi w} f(x) e^{-ix\nu/w}\, dx, \qquad \nu \in \mathbb{R},$$

exist. When ν takes integer values the $\{c_\nu\}$ are called *Fourier coefficients* of f. Then f generates a *formal Fourier series* :

$$f(x) \sim \frac{1}{\sqrt{2\pi w}} \sum_{n \in \mathbb{Z}} c_n e^{ixn/w}. \tag{2.1.1}$$

We shall also need the expansion of a function f in the complex conjugates of the exponentials in this formal Fourier series; that is,

$$f(x) \sim \frac{1}{\sqrt{2\pi w}} \sum_{n \in \mathbb{Z}} d_n e^{-ixn/w}, \tag{2.1.2}$$

where

$$d_n = \frac{1}{\sqrt{2\pi w}} \int_{\pi w}^{\pi w} f(x) e^{ixn/w}\, dx. \tag{2.1.3}$$

Either form will be called the Fourier series.

Theorem 2.5 *Equality replaces " $\sim$ " when an integrable function f and its formal Fourier series are multiplied by any function of bounded variation and integrated over any bounded interval.*

Theorem 2.6 (Riemann–Lebesgue) *If f is integrable over an interval of periodicity, then c_ν has the property: $c_\nu \to 0$ as $|\nu| \to \infty$. If f is of bounded variation, $c_\nu = O\left(\frac{1}{\nu}\right)$, $|\nu| \to \infty$.*

2.2 The Fourier transform

We start with the definition of the Fourier transform in N dimensions; this will be needed in Chapter 14. Mostly however, it is the one-dimensional theory that is needed and the remainder of our review will revert to that case. A few more details about Fourier analysis in N dimensions will be found in §14.1, but it should be noted here that virtually all the results on Fourier analysis in the previous section and in this one have straightforward extensions to the multi-dimensional case.

Let x and t be N-dimensional multi-variables, and let $x \cdot t$ denote the usual scalar product.

Definition 2.7 *The Fourier transform of a complex valued function $f(t)$, $t \in \mathbb{R}^N$, is defined formally by*

$$(\mathcal{F}f)(x) = f^\wedge(x) = \frac{1}{(2\pi)^{N/2}} \int_{\mathbb{R}^N} f(t)e^{-ix\cdot t}\, dt.$$

At the moment this definition is formal only; the integral must be interpreted according to the nature of f. For example, we have:

Theorem 2.8 *If $f \in L^1(\mathbb{R})$ the Fourier transform exists in the sense:*

$$f^\wedge(x) = \frac{1}{\sqrt{2\pi}} \lim_{A\to\infty} \int_{-A}^{A} f(t)e^{-ixt}\, dt.$$

Definition 2.9 *$C_0(\mathbb{R})$ denotes the class of functions f which are continuous over $\mathbb{R}$, and for which $f(t) \to 0$ as $|t| \to \infty$.*

Theorem 2.10 *$\mathcal{F} : f \mapsto f^\wedge$ defines a bounded linear transformation from $L^1(\mathbb{R})$ into $C_0 \subset L^\infty(\mathbb{R})$.* (Stein and Weiss 1971, p. 2).

Theorem 2.11 *If both f and $f^\wedge$ belong to $L^1(\mathbb{R})$, the Fourier inversion formula holds:*

$$f(t) = \left(\mathcal{F}^{-1}f^\wedge\right)(t) = \frac{1}{\sqrt{2\pi}} \int_{\mathbb{R}} f^\wedge(x)e^{ixt}\, dx, \qquad a.e.$$

(Stein and Weiss 1971, p. 11).

Definition 2.12 *The Fourier transform on $L^2(\mathbb{R})$ is taken to be the unique extension to all of $L^2(\mathbb{R})$ of $\mathcal{F}$ defined on the dense subset $L^1(\mathbb{R}) \cap L^2(\mathbb{R})$ of*

$L^2(\mathbb{R})$. *Equivalently, the Fourier transform of* $f \in L^2(\mathbb{R})$ *is defined as the limit in* L^2*-norm of the symmetric partial integrals*

$$\frac{1}{\sqrt{2\pi}} \int_{|t| \leq A} f(x) e^{-ixt}\, dx,$$

as $A \to \infty$.

Theorem 2.13 (Plancherel) $\mathcal{F}: f \mapsto f^\wedge$ *is a bounded linear transformation of* $L^2(\mathbb{R})$ *onto itself, such that* $\|f\|_2 = \|f^\wedge\|_2$ *for every* f *belonging to* $L^2(\mathbb{R})$.

The Fourier transform $\mathcal{F}$ is, in particular, *isometric* on $L^2(\mathbb{R})$ in the sense that it preserves $L^2(\mathbb{R})$-norms.

We shall usually refer to a transformation whose domain and range belong to the same Hilbert space as an *operator* on that space. A bounded, linear, invertible and norm preserving operator on a Hilbert space is called a *unitary operator*; so the Fourier transform is an example of a unitary operator on $L^2(\mathbb{R})$.

Furthermore, Fourier inversion is obtained in L^2 by taking $(\mathcal{F}^{-1}g)(t) := (\mathcal{F}g)(-t)$.

Definition 2.14 *Let* $f \in L^p(\mathbb{R})$, $1 < p \leq 2$. *The Fourier transform* $f^\wedge$ *of* $f \in L^p(\mathbb{R})$ *is defined as the limit in the norm of* L^q *of the symmetric partial integrals* $(2\pi)^{-\frac{1}{2}} \int_{|t| \leq A} f(x) e^{-ixt}\, dx$, *as* $A \to \infty$.

Reciprocally, Fourier inversion holds in the following sense:

Theorem 2.15 *Let* $f \in L^p(\mathbb{R})$, $1 < p \leq 2$. *Then*

$$f(t) = \lim_{A \to \infty} \frac{1}{\sqrt{2\pi}} \int_{-A}^{A} \left(1 - \frac{|x|}{A}\right) f^\wedge(x) e^{ixt}\, dx, \qquad a.e.$$

One says that the Fourier inversion integral is Cesàro summable to f almost everywhere (Butzer and Nessel 1971, p. 213).

When $p > 2$ the Fourier transform $f^\wedge$ of $f \in L^p(\mathbb{R})$ must be interpreted in the sense of distributions (Problem 2.4.)

Theorem 2.16 *The mapping* $\mathcal{F}: f \mapsto f^\wedge$ *is a bounded linear transformation of* $L^p(\mathbb{R})$ *into* $L^q(\mathbb{R})$, $1 < p \leq 2$.

The boundedness comes from the Hausdorff–Young inequality (Theorem 2.17 below). This inequality and two more of the basic inequalities of Fourier analysis are recorded here in the next three theorems; all three are associated with the names of Hausdorff and W.H. Young. Explicit values for the best possible constants in these inequalities are not listed since they do not concern us directly; they can be found, together with much further interesting material, in the historical study by Butzer (1994), and in the expository account by Stein (1976 §6,7).

Theorem 2.17 (Hausdorff–Young inequality for transforms)
Let $f \in L^p(\mathbb{R})$, $1 \le p \le 2$. Then there exists a constant C_p, depending only on p, such that

$$\|f^\wedge\|_q \le C_p \|f\|_p.$$

There is a companion inequality for Fourier coefficients.

Theorem 2.18 (Hausdorff–Young inequality for coefficients)
Suppose that $f \in L^p(-\sigma, \sigma)$, $1 < p \le 2$, and let $\{c_n\}$ be the Fourier coefficients of f. Then there exists a constant $D_{\sigma,p}$, depending only on σ and p, such that

$$\left\{ \sum_{n \in \mathbb{Z}} |c_n|^q \right\}^{1/q} \le D_{\sigma,p} \left\{ \int_{-\sigma}^{\sigma} |f(u)|^p \, du \right\}^{1/p}.$$

Definition 2.19 *The convolution $f * g$ of f with g is defined to be*

$$(f * g)(t) = \frac{1}{\sqrt{2\pi}} \int_{\mathbb{R}} f(v)g(t - v) \, dv.$$

Whenever this integral exists we evidently have $f*g = g*f$. Another standard inequality in Fourier analysis tells us when we can expect a convolution to exist. It is as follows:

Theorem 2.20 (Young's convolution inequality) *Suppose that $f \in L^r(\mathbb{R})$, $g \in L^s(\mathbb{R})$. Then provided that*

$$\frac{1}{p} = \frac{1}{r} + \frac{1}{s} - 1, \qquad 1 \le p, r, s \le \infty,$$

we have $f * g \in L^p(\mathbb{R})$ and there exists a constant $E_{p,r,s}$ such that

$$\|f * g\|_p \le E_{p,r,s} \|f\|_r \|g\|_s.$$

An account of this inequality, with further references, can be found in Stein (1976 §7).

2.3   Poisson's summation formula

A formula that links f with its Fourier transform $f^\wedge$ in a fundamental way is Poisson's summation formula[1]

---

[1]Although this formula is named after Poisson, a basic form appears rather early in in the history of Fourier analysis in a note by Gauss (without proof) written some time between 1799 and 1813 (see p. 88 of *Carl Friedrich Gauss Werke*, Band 8, Königl. Gesellschaft Wiss. Gottingen, Teubner, Leipzig, 1900).

$$\sum_{n\in\mathbb{Z}} f\left(\frac{1}{w}(n+t)\right) \sim w\sqrt{2\pi}\sum_{n\in\mathbb{Z}} f^\wedge(2\pi wn)e^{2\pi int}. \qquad (2.3.1)$$

Initially the symbol " $\sim$ " means only that the left- and right-hand sides are ways of forming periodic functions from $f^\wedge$, with period 1. In the following theorem we narrow matters to the point where " $\sim$ " will mean the presence of a formal Fourier expansion.

Theorem 2.21 *Let $f \in L^1(\mathbb{R})$. Then the left-hand side of (2.3.1) converges in the norm of $L^1(-1/2,1/2)$ to a function whose formal Fourier series is the right-hand side of (2.3.1).*

Proof For the first part of the proof we have

$$\int_{-1/2}^{1/2}\left|\sum_{n\in\mathbb{Z}} f\left(\frac{1}{w}(n+t)\right)\right| dt \le \int_{-1/2}^{1/2}\sum_{n\in\mathbb{Z}}\left|f\left(\frac{1}{w}(n+t)\right)\right| dt. \qquad (2.3.2)$$

Now we can interchange the order of integration and summation on the right-hand side of (2.3.2) if the expression exists with either ordering. In fact it is convenient to show existence in the opposite order to that shown. Indeed,

$$\sum_{n\in\mathbb{Z}}\int_{-1/2}^{1/2}\left|f\left(\frac{1}{w}(n+t)\right)\right| dt = \sum_{n\in\mathbb{Z}}\int_{n-1/2}^{n+1/2}|f(v/w)|\,dv$$

$$= \int_{\mathbb{R}}|f(v/w)|\,dv < \infty.$$

This gives the required L^1 convergence. We now use the same idea on interchanging sum and integral to prove the second part; thus the kth Fourier coefficient of the left-hand side of (2.3.1) is

$$\int_{-1/2}^{1/2}\sum_{n\in\mathbb{Z}} f\left(\frac{1}{w}(n+t)\right)e^{-2\pi ikt}\,dx = \sum_{n\in\mathbb{Z}}\int_{n-1/2}^{n+1/2} f(v/w)e^{-2\pi ik(v-n)}\,dv$$

$$= \int_{\mathbb{R}} f(v/w)e^{-2\pi ivk}\,dv = w\int_{\mathbb{R}} f(u)e^{-2\pi iwuk}\,du = w\sqrt{2\pi}f^\wedge(2\pi wk),$$

the last equality follows from the L^1 definition of the Fourier transform. This verifies the right-hand side of (2.3.1) and completes the proof.

A strengthened form of Poisson summation will be obtained in the proof of Theorem 6.11, and a different formulation, one that is suitable for the study of aliasing errors, will be derived in §12.1.

2.4   Tempered distributions — some basic facts

Some very brief facts about distributions are included here, mainly as background for the appreciation of Theorem 6.27. These basic ideas will be built upon in H–S.

Definition 2.22 *The Schwartz space S of "functions of rapid descent at infinity" consists of those functions f such that $f \in C^\infty(\mathbb{R})$ (that is, f and its derivatives of all orders are continuous on $\mathbb{R}$), and*

$$\sup_{x \in \mathbb{R}} \left| x^\alpha f^{(\beta)}(x) \right| < \infty, \qquad \alpha \in \mathbb{N}_0, \quad \beta \in \mathbb{N}_0.$$

For example, e^{-ax^2}, $a > 0$, belongs to S but $e^{-a|x|}$ does not, since it is not differentiable at the origin.

We now list some useful properties of S:

1. S contains all C^∞ functions of compact support (the *support* of a C^∞ function f is the closure of the set $\{x : f(x) \neq 0\}$).

(Functions of this kind need some care in their construction. As an example we can take

$$f(x) = \begin{cases} e^{-2/(1-x^2)}; & |x| \leq 1 \\ 0 & |x| > 1. \end{cases}$$

This is sometimes called a "bump function".)

2. S is a dense subspace of $C_0(\mathbb{R})$ and of every $L^p(\mathbb{R})$, $1 \leq p < \infty$.
3. The Fourier transform $\mathcal{F}$ is surjective (one-to-one and onto) on S.

Definition 2.23 *The collection S' of all bounded linear functionals on S is called the space of tempered distributions.*

There are many ways of generating tempered distributions. For example, an ordinary function $f \in L^p(\mathbb{R})$ generates a bounded linear functional l on S, that is, a tempered distribution, by taking

$$l(\varphi) = \int_{\mathbb{R}} f(x)\varphi(x)\,dx, \qquad \varphi \in S.$$

Again, $l(\varphi) = (\mathcal{D}^\beta \varphi)(x_0)$ defines a tempered distribution. A special case of this is *Dirac's delta*: $l(\varphi) = \varphi(0)$.

Definition 2.24 *The Fourier transform $l^\wedge$ of a tempered distribution l is the bounded linear functional given by*

$$l^\wedge(\varphi) = l(\varphi^\wedge).$$

PROBLEMS

2.1 Show that if $f \in L^1(\mathbb{R})$ and f is bounded over $\mathbb{R}$, then $f \in L^2(\mathbb{R})$.

2.2 Show that if $f \in L^p(\mathbb{R})$, and supp $f^\wedge = [-\sigma, \sigma]$, $1 \leq p \leq 2$, then f is equal to a continuous function almost everywhere.

2.3 Show that if $f \in L^p(\mathbb{R})$, $1 \leq p \leq 2$, then its Fourier transform in the distributional sense coincides with its transform in the ordinary sense.

2.4 Show that when $f \in L^p(\mathbb{R})$, $p > 2$, its Fourier transform exists in the sense of tempered distributions.

2.5 Show that, as a consequence of Definition 2.24, $(\mathbf{1}^\wedge)(x) = \sqrt{2\pi}\delta(x)$, where $\mathbf{1}$ denotes the function which is constantly 1 on $\mathbb{R}$.

3

HILBERT SPACES, BASES AND FRAMES

Hilbert spaces of various types provide the natural setting for much of the sampling theory that we shall be studying.

The reader is assumed to be familiar with the foundations of Hilbert and Banach spaces. We review here some of the important facts, mostly concerning Hilbert spaces, that will be needed throughout the text. Of particular importance is the rather more specialized topic of bases in Hilbert space. An elementary account can be found in the book by Higgins (1977). A very readable introduction to the theory of bases is in the book by Young (1980); the book of Marti (1969) has a more advanced treatment, while a vastly comprehensive treatment is to be found in the book of Singer (1970).

Apart from the last section this chapter is intended as a review only, and almost no proofs are included. However, ample references to the texts mentioned above are given.

3.1 Bases for Banach and Hilbert spaces

Let B denote a seperable Banach space over the field $\mathbb{C}$ of complex numbers. We shall always denote the identity element of B by θ. A subset $\Phi \subset B$ is called *complete* in B if the closed linear span of Φ is all of B.

The sampling series which form our main focus of interest will frequently be seen as expansions for members of various Hilbert and Banach spaces, and we therefore need expansion properties which are stronger than those afforded by sets which are merely complete. Consequently we shall review some important properties of bases in this section and the next. Throughout this chapter X will denote an indexing set; it will always be $\mathbb{Z}$, $\mathbb{Z}^N$ or one of their subsets.

We shall not usually make a distinction between a sequence (h_n) of members of B and a subset $\{h_n\} \subset B$; the latter notation is used on the understanding that the h_n are ordered by the natural ordering of the subscript when the context requires it.

Definition 3.1 *A basis for the strong topology for B is a subset $\{h_n\} \subset B$, $n \in X$, which determines a unique collection of coefficient functionals $\{c_n\}$ such that*

$$f = \sum_{n \in X} c_n(f) h_n$$

in the strong topology, for every f belonging to B.

Let H denote a separable Hilbert space, with inner product denoted by $\langle .,. \rangle$, and norm $\|.\|$ (a subscript will sometimes be attatched to these symbols for emphasis). Throughout this chapter and most of the subsequent text H will be infinite-dimensional; however, in Chapter 4 we shall need some of the results collected here in the finite-dimensional case. This usage is valid by a trivial reduction.

Essential tools for working in Hilbert spaces are the *Cauchy–Schwarz inequality* $|\langle f,g\rangle| \leq \|f\|\,\|g\|$ and the *triangle inequalities* $|\,\|f\| - \|g\|\,| \leq \|f+g\| \leq \|f\| + \|g\|$, which hold for all $f, g \in$ H.

Hilbert spaces are unique among classes of abstract spaces in that they alone are guaranteed to possess a *Schauder basis*, a basis for which every c_n is a bounded linear functional. All bases for Banach spaces are Schauder bases (Marti 1969, p. 29), and since we shall not consider any other kind of space here the name Schauder will be omitted when refering to bases.

Two members h_1 and h_2 of a Hilbert space H are said to be *orthogonal* if $\langle h_1, h_2 \rangle = 0$; this is denoted by $h_1 \perp h_2$. A subset $\{h_n\}$, $n \in$ X, in a Hilbert space H is said to be an *orthogonal subset* if $h_n \perp h_m$, $n \neq m$. It is said to be *orthonormal* if $\langle h_n, h_m \rangle = \delta_{nm}$, $n, m \in$ X, where δ_{nm} is Kronecker's delta symbol.

Theorem 3.2 *A Hilbert space H contains an orthonormal basis, that is, a subset $\{h_n\}$ that is both orthonormal and a basis. For such bases we have $c_n(f) = \langle f, h_n \rangle$, $n \in$ X. We have the General Parseval Relation*

$$\langle f,g\rangle = \sum_{n\in X} c_n(f)\overline{c_n(g)} \tag{3.1.1}$$

whenever $f, g \in$ H. The special case $f = g$ gives the ordinary Parseval Relation

$$\|f\|^2 = \sum_{n\in X} |c_n(f)|^2.$$

(Young 1980, p. 7).

An important implication of the Parseval Relation is that $c_n(f) \to 0$ as $n \to \infty$. This should be compared with the Theorem 2.6 of Riemann–Lebesgue.

In a separable Hilbert space an orthogonal set is a basis if and only if it is *complete*; that is, the class of all linear combinations of members of the set is dense in the space. The property: $f \perp \varphi_n$, $n \in$ X, implies $f = \theta$, is equivalent to the completeness of $\{\varphi_n\}$ (see, for example, Higgins 1977, p. 15). Here we shall use the word *complete* for both these equivalent properties, but there are several other words in current use, e.g. fundamental, total, closed, etc.

Theorem 3.3 (Riesz–Fischer) *Let $\{h_n\}$, $n \in$ X, be an orthonormal subset of a Hilbert space H, and (a_n) a sequence of scalars. Then there exists an $f \in$ H such that $\sum_{n\in X} a_n h_n$ converges to f if and only if $\sum_{n\in X} |a_n|^2 < \infty$. Under either condition $a_n = \langle f, h_n \rangle$. If, furthermore, $\{h_n\}$ is complete then there cannot exist a $g \in$ H, distinct from f, for which $a_n = \langle g, h_n \rangle$ (see, for example, Higgins 1977 p. 16).*

The Hilbert space l^2 of all square summable sequences $\{(z_1, \dots)\}$ of complex numbers is typical, in the sense that all separable infinite-dimensional Hilbert spaces are isometrically isomorphic to it under the mapping which takes $f \in H$ to $(\langle f, h_1 \rangle, \dots) \in l^2$, where $\{h_n\}$ is an orthonormal basis for H (Marti 1969, p. 80).

For example, the standard orthonormal basis for l^2 is $\{e_n\}$, $n \in \mathbb{N}$, where $e_n = (0, \dots 0, 1, 0, \dots)$, the 1 appearing as the nth co-ordinate; and if I is any interval of length 2σ, then $\{h_n\}$, $n \in \mathbb{Z}$ is an orthonormal basis for $L^2(I)$, where $h_n(x) = e^{in\pi x/\sigma}/\sqrt{2\sigma}$. This result is clearly translation invariant, so that $\{e^{in\pi x/\sigma} e^{iax}/\sqrt{2\sigma}\}$, $n \in \mathbb{N}$, is also an orthonormal basis for $L^2(I)$, for every real a.

Definition 3.4 *A biorthonormal system in a Hilbert space H consists of two subsets $\{\varphi_n\}$ and $\{\varphi_n^\star\}$ which both belong to H, the collection being usually written $\{\varphi_n, \varphi_n^\star\}$, such that $\langle \varphi_n, \varphi_m^\star \rangle = \delta_{nm}$.*

Theorem 3.5 *If $\{\varphi_n\}$, $n \in X$, is a basis for H, then it determines a unique subset $\{\varphi_n^\star\}$ in H such that $\{\varphi_n, \varphi_n^\star\}$ is a biorthonormal system; $c_n(f) = \langle f, \varphi_n^\star \rangle$, and $\{\varphi_n^\star\}$ is also a basis for H.* (Marti 1969, pp. 30–33).

Definition 3.6 *When $\{\varphi_n\}$, $n \in X$, is a basis for H, the biorthonormal system $\{\varphi_n, \varphi_n^\star\}$ of Theorem 3.5 constitutes a pair of bases for H, in which each basis is said to be the dual of the other.*

3.2   Riesz bases and unconditional bases

If H and H$'$ are separable Hilbert spaces, the two subsets $\{\varphi_n\} \subset$ H and $\{\psi_n\} \subset$ H$'$ are said to be *fully equivalent* (Singer 1970, p. 69) if there is a bounded linear bijection (an isomorphism in the sense of Banach), T say, of H onto H$'$ which, in particular, maps φ_n to ψ_n for every n. The name *equivalent* is justified because, by Banach's inversion theorem (sometimes called the "bounded inverse theorem", as in Lorch 1962, p. 38), our isomorphism has an inverse, S say, which is also an isomorphism, and in particular maps ψ_n to φ_n for every n.

Definition 3.7 *A basis $\{\varphi_n\}$ for H is called a Riesz basis if it is fully equivalent to an orthonormal basis for H$'$.*

Evidently an orthonormal basis for H is already a Riesz basis.

Theorem 3.8 *The transformation S induces a transformation $S^* : H \mapsto H'$ that satisfies*

$$\langle Sh', h \rangle_H = \langle h', S^\star h \rangle_{H'}.$$

It is called the adjoint transformation of S. (Lorch 1962, p. 36).

Now let $\{e_n\}$ be an orthonormal basis for H$'$. Its image in H by S is a Riesz basis $\{\varphi_n\}$ for H. We shall see that its biorthonormal basis is closely related to the adjoint transformation $S^\star$. Indeed, let us define $\varphi_n^\star$ by $(S^\star)^{-1} e_n = \varphi_n^\star$. Then

$$\langle \varphi_n, \varphi_m^\star \rangle_H = \langle Se_n, (S^\star)^{-1} e_m \rangle_H = \langle e_n, S^\star (S^\star)^{-1} e_m \rangle_{H'} = \delta_{nm}.$$

Hence $\{\varphi_n, \varphi_n^\star\}$ is a biorthonormal system in H, and so $\{\varphi_n^\star\}$ is the basis biorthonormal to $\{\varphi_n\}$.

Definition 3.9 *A basis $\{\varphi_n\}$ for H is called an unconditional basis if, for every $f \in H$, the expansion $\sum c_n(f)\varphi_n$ still converges to f after any permutation of its terms.*

An orthonormal basis for H is unconditional (Young 1980, p. 13). Now suppose that $\{\varphi_n\}$ is a Riesz basis for H. We have the following:

PROPERTIES OF RIESZ BASES.

1. The biorthonormal basis, or *dual basis* $\{\varphi_n^\star\}$, is also a Riesz basis for H; (Young 1980, p. 36; Marti 1969, Ch. III, §2.).
2. There exist positive constants a and b such that

$$a \sum_n |c_n|^2 \leq \left\| \sum_n c_n\varphi_n \right\|^2 \leq b \sum_n |c_n|^2,$$

   for all finite sequences of scalars $\{c_n\}$. (Young 1980, p. 32).
3. A Riesz basis is in particular a Riesz–Fischer sequence; that is, the moment problem
$$\langle f, \varphi_n \rangle = a_n \qquad \text{for every } n$$
   has a solution in H whenever $(a_n) \in l^2$ (Young 1980, pp. 154–155).
4. If a subset in H and another in H′ are fully equivalent, then either is a basis if the other is (Singer 1970, p. 78). It follows that if a subset in H is fully equivalent to an orthonormal basis for H′, it is a Riesz basis for H.
5. A Riesz basis is an unconditional basis for H. Equivalently, it is a *subseries convergent basis* (Marti 1969, p. 84; Singer 1970, p. 458); that is, a basis such that for every $f \in H$, and every subsequence $(n_k) \subset X$, $k \in N$, $\sum_{k \in N} c_{n_k}(f)\varphi_{n_k}$ converges.

It is very convenient to invoke property 4 when we need to show that a given subset $\{\varphi_n\}$ is a Riesz basis for a Hilbert space H. It means that we do not need to show, as Definition 3.7 might suggest, that $\{\varphi_n\}$ is a basis for H as well as to prove its full equivalence to an orthogonal basis for H′; in fact the second condition suffices.

The notion of unconditional basis is slightly weaker than that of Riesz basis. On the one hand, we have seen (in property 5 above) that every Riesz basis is unconditional. On the other hand we have:

Theorem 3.10 *An unconditional basis $\{\varphi_n\}$ that also satisfies*

$$\inf_n \|\varphi_n\| > 0 \quad and \quad \sup_n \|\varphi_n\| < \infty$$

(this condition can be written $\|\varphi_n\| \asymp 1$ in the notation introduced in §3.3 below) is a Riesz basis.

This is a theorem of Köthe and dates from 1936. An account of unconditional bases that is very appropriate to our present theme can be found in Hruščev *et al.* (1981).

The following characterization of unconditional bases makes an interesting, and for our purposes important, contrast to that of Riesz bases.

Theorem 3.11 *A subset $\{\varphi_n\}$ of a Hilbert space H is an unconditional basis if:*
1. $\{\varphi_n\}$ spans H.
2. There exist positive constants α and β such that

$$\alpha \sum_n |c_n|^2 \|\varphi_n\|^2 \leq \left\| \sum_n c_n \varphi_n \right\|^2 \leq \beta \sum_n |c_n|^2 \|\varphi_n\|^2 ,$$

for all finite sequences of scalars $\{c_n\}$. (Hruščev et al. 1981, p. 216).

In Chapter 10 we shall find that it is the Riesz bases that are the more useful, particularly in the context of signal theory, because we can associate a notion of stable sampling with them, and this is not the case for bases that are merely unconditional.

Now the Fourier transform is a unitary operator on $L^2(\mathbb{R})$ and thus induces an isomorphism between PW_B and $L^2(B)$. This is an important duality principle; for example, we shall make repeated use of the following consequence of it.

Theorem 3.12 *Under Fourier transformation, the pre-image of a Riesz basis for $L^2(B)$ is a Riesz basis for PW_B.*

3.3   Frames

First a useful piece of notation will be introduced. Suppose that l_1 and l_2 are linear functionals such that $l_1(f) \geq 0$ and $l_2(f) \geq 0$ for every f belonging to a certain domain F. Then

$$l_1(f) \asymp l_2(f)$$

is a succinct way of expressing the fact that there exist positive constants c and C, independent of f, such that

$$c l_2(f) \leq l_1(f) \leq C l_2(f).$$

Such a condition implies that the quotient of $l_1(f)$ by $l_2(f)$ is bounded away from 0 and also bounded from above.

Definition 3.13 *A subset $\{h_n\}$, $n \in \mathbb{X}$, in a separable Hilbert space H is called a frame if, for every $f \in H$,*

$$\sum_{n \in \mathbb{X}} |\langle f, h_n \rangle|^2 \asymp \|f\|^2.$$

The implied constants c and C are called the frame bounds. If $c = C$ the frame is said to be tight.

It follows that a frame is a complete set in H, since if f is orthogonal to every h_n then it must be null. However, frames are not in general bases; rather they are "over-complete" sets in the sense that any one member belongs to the closed linear span of the others. This kind of redundancy means that their associated coefficient functionals are not unique (but see also Problem 3.5).

If a frame has the property that the removal of any one member leaves an incomplete set, then the frame is said to be *exact*.[1]

It is interesting to note that a frame can be tight and have unit frame bounds, yet not be exact (as in Example 10.17).

Theorem 3.14 *A subset in a Hilbert space H is an exact frame if and only if it is a Riesz basis.* (Young 1980, p. 188).

Definition 3.15 *If $\{h_n\}$ is a frame for H, the operator $\mathcal{S}$ given by*

$$\mathcal{S}: f \longmapsto \sum_{n \in \mathbb{X}} \langle f, h_n \rangle h_n$$

is called the frame operator.

Theorem 3.16 *$\mathcal{S}$ is an isomorphism of H onto H.*

Put $g_n = \mathcal{S}^{-1} h_n$; then (g_n) is called the *dual frame* to $\{h_n\}$.

Theorem 3.17 *The subset (g_n) is indeed a frame and satisfies the dual frame condition*

$$\sum_{n \in \mathbb{X}} |\langle f, g_n \rangle|^2 \asymp \|f\|^2$$

for every $f \in H$, with bounds $1/C$ and $1/c$, where c and C are the bounds in Definition 3.13.

The coefficient functionals associated with a frame are not unique, but there are two "canonical" frame expansions in a Hilbert space H.

Theorem 3.18 *For every $f \in H$*

$$f = \sum_{n \in \mathbb{X}} \langle f, h_n \rangle g_n = \sum_{n \in \mathbb{X}} \langle f, g_n \rangle h_n.$$

Suppose that $\Lambda = \{\lambda_n\}$, $n \in \mathbb{X}$, is a sequence of real or complex numbers for which $\{e^{i\lambda_n x}\}$ forms a frame in $L^2(-\pi w, \pi w)$. These "non-harmonic Fourier frames" are the only kind of frame that we need to consider here. Real sequences Λ which generate such frames have been completely described by Jaffard (1991), a result quoted in Theorem 10.16.

---

[1] Caution! This does not mean that every frame contains an exact frame as a subset (Seip 1995b).

3.4 Reproducing kernel Hilbert spaces

A good introduction to this topic can be found in the book of Saitoh (1988).

Let H be a separable Hilbert space whose members are functions defined on a given domain D. Suppose there exists a "kernel" k defined on $D \times D$ such that two conditions hold:

(a) As a function of x, $k(x, y)$ belongs to H for each $y \in D$.
(b) For every $f \in H$,

$$f(y) = \langle f, k(\cdot, y) \rangle, \qquad y \in D,$$

the *reproducing formula*.

Definition 3.19 *A kernel k satisfying the two conditions above is called a reproducing kernel ("rk" for short). A Hilbert space possessing a reproducing kernel is called a reproducing kernel Hilbert space ("rk space" for short).*

PROPERTIES OF rk SPACES.

1. A Hilbert function space H is an rk space if and only if "point evaluation", $l_y f = f(y)$, is a bounded linear functional on H.
2. A Hilbert space can possess at most one rk.
3. If $\{\varphi_n\}$, $n \in X$, is an orthonormal basis for an rk space H, then in the sense of strong convergence the rk is given by

$$k(x, y) = \sum_{n \in X} \varphi_n(x) \overline{\varphi_n(y)}.$$

It follows that

$$k(x, y) = \overline{k(y, x)}.$$

3'. If $\{\varphi_n\}$, $n \in X$, is a basis for an rk space H with biorthonormal basis $\{\varphi_n^\star\}$, $n \in X$, then in the sense of strong convergence the rk is given by

$$k(x, y) = \sum_{n \in X} \varphi_n(x) \overline{\varphi_n^\star(y)}.$$

4. The convergence of a sequence in an rk space implies its pointwise convergence over D to the same sum; the convergence is uniform over any subset of D for which $k(x, x)$ is bounded.

While rk spaces do exist, not every Hilbert space is one. For example, since their members are equivalence classes rather than functions, the point evaluation functional is not properly defined on L^2 spaces, and property 1 above fails. On the other hand we have:

Example 3.20 The Hilbert space l^2 is an rk space, since if $(z_n) \in l^2$, then $|z_m|^2 \le \sum_{n \in X} |z_n|^2$, $m \in \mathbb{N}$. Hence $|z_m| \le \|z\|_{l^2}$, and "evaluation" is bounded. The rk for l^2 is evidently δ_{mn}, $m, n \in \mathbb{N}$, since $z_m = \sum_{i \in X} z_i \delta_{im} = \langle z, \delta. \, m \rangle$.

We noted in §3.1 that every Hilbert space is isometrically isomorphic to l^2. The last example shows that the property of being an rk space is not an invariant of such mappings.

3.5   Direct sums of Hilbert spaces

Let H and H′ be separable Hilbert spaces. The set of all ordered pairs (h, h') with $h \in H$ and $h' \in H'$ is a vector space under addition and scalar multiplication defined by

$$(h_1, h_1') + (h_2, h_2') := (h_1 + h_2, h_1' + h_2'),$$
$$\alpha(h, h') := (\alpha h, \alpha h').$$

On this space we adopt the inner product defined by

$$\langle (h_1, h_1'), (h_2, h_2') \rangle = \langle h_1, h_2 \rangle_H + \langle h_1', h_2' \rangle_{H'}.$$

A norm is then defined by

$$\|(h, h')\| := \{\|h\|_H^2 + \|h'\|_{H'}^2\}^{\frac{1}{2}}.$$

With these definitions of norm and inner product our vector space becomes a Hilbert space called the *direct sum* of H and H′ and denoted by $H \oplus H'$ (it is sometimes called an *external direct sum*).

Suppose that $S \subset H$, and that for some $h \in H$ we have $h \perp k$ whenever $k \in S$. Then h is said to be *orthogonal to* S, and this is denoted by $h \perp S$. Two subsets H and K of a Hilbert space H are said to be orthogonal (denoted by $H \perp K$) if $h \perp K$ for every $h \in H$.

Let M be a subspace of H. Then

$$M^\perp := \{h \in H : h \perp M\}$$

is called the *orthogonal complement of* M. We have $M \perp M^\perp$, and $H = M \oplus M^\perp$. This is sometimes called an *internal direct sum*; the form given here is also called an *orthogonal direct sum*. The subspaces M and $M^\perp$ are called a *pair of complementary subspaces*.

The definition of a direct sum space clearly extends naturally to any finite number of summands.

We shall need to extend the ideas of bases to direct sum spaces, as follows.

Theorem 3.21 *Let* $\{\varphi_i\}, i \in \mathbb{X}$, *be a basis for* H *and* $\{\varphi_j'\}, j \in \mathbb{X}$, *be a basis for* H′. *Then* $\{\psi_n\}, n \in \mathbb{X}$, *defined by*

$$\begin{cases} \psi_{2i} = (\varphi_i, \theta) \\ \psi_{2j+1} = (\theta, \varphi_j'), \end{cases}$$

is a basis for $H \oplus H'$. (Singer 1970, p. 28).

Of course, if both $\{\varphi_i\}$ and $\{\varphi'_j\}$ are orthonormal then so is $\{\psi_n\}$. This theorem extends in a natural way to direct sums with any finite number of summands.

3.6   Sampling and reproducing kernels

Preliminary to our study of sampling series in later chapters we now prove two lemmas giving some facts about bases with sampling properties in rk Hilbert spaces.

Let H be an rk Hilbert space of functions defined on a domain D, which for our purposes will be a Euclidean space, or a subset of one. Let H have the reproducing kernel $k(x,y)$ and let (λ_n), $n \in \mathbf{X}$, be a sequence of points belonging to D.

Definition 3.22 *A sequence of functions $\{\varphi_n\}$, $n \in \mathbf{X}$, belonging to H is said to have the sampling property with respect to (λ_n) if $\varphi_n(\lambda_m) = \delta_{nm}$, $n, m \in \mathbf{X}$.*

Lemma 3.23 *Let (λ_n) be such that $\{k(\cdot, \lambda_n)\}$ is a basis for H and let $\{S_n^\star\}$ be the dual basis. Then, for every $f \in H$, we have the sampling formula*

$$f = \sum_{n\in\mathbf{X}} f(\lambda_n)S_n^\star, \qquad (3.6.1)$$

with convergence in norm. If, in particular, $\{k(\cdot, \lambda_n)\}$ is an orthogonal basis for H, (3.6.1) becomes

$$f = \sum_{n\in\mathbf{X}} f(\lambda_n)k(\cdot, \lambda_n)/k(\lambda_n, \lambda_n).$$

Proof If f is expanded in the set $\{S_n^\star\}$, we obtain

$$f = \sum_{n\in\mathbf{X}} \langle f, k(\cdot, \lambda_n)\rangle S_n^\star = \sum_{n\in\mathbf{X}} f(\lambda_n)S_n^\star.$$

The second part is now immediate, since

$$\|k(\cdot, \lambda_n)\|^{-1} = \langle k(\cdot, \lambda_n), k(\cdot, \lambda_n)\rangle^{-1/2} = k(\lambda_n, \lambda_n)^{-1/2}$$

are the appropriate normalizing factors.

By way of a converse to this lemma, we have:

Lemma 3.24 *Let $\{\varphi_n\}$, $n \in \mathbf{X}$, be a basis for H such that **either** (a) we have*

$$f = \sum_{n\in\mathbf{X}} f(\lambda_n)\varphi_n^\star, \qquad f \in H;$$

***or** (b) the set $\{\varphi_n\}$ has the sampling property with respect to $\{\lambda_n\}$. Then $\varphi_n = k(\cdot, \lambda_n)$.*

Proof For part (a) we find that, since $\{\varphi_n\}$ is a basis, $f(\lambda_n) = \langle f, \varphi_n \rangle$. Hence, by uniqueness of k, $\varphi_n = k(\cdot, \lambda_n)$.

For part (b), from the expansion $f = \sum a_n \varphi_n^*$ we find from the sampling property that $a_n = f(\lambda_n)$. The result now follows as in part (a).

It is interesting to note that any basis $\{\varphi_n\}$ for an rk Hilbert space that has the sampling property with respect to a sequence $(\lambda_n) \subset D$ must arise from the reproducing kernel, so that $\varphi_n = k(\cdot, \lambda_n)$, $n \in \mathbb{X}$.

PROBLEMS

3.1 Suppose that a set is *complete* in a Hilbert space. Show that if it has the additional property of being *orthogonal* it is a basis for the space; but that if instead of orthogonality it has the additional property of being merely *linearly independent* (that is, none of its members is contained in the closed linear span of the others), it may fail to be a basis.

In this sense, orthogonality is a stronger property than linear independence.

3.2 Prove Theorem 3.2 (Zorn's Lemma or one of its equivalents will be needed).

3.3 Let H be a Hilbert space that is algebraically closed under complex conjugation. Show that $\{\varphi_n\}$ is a basis if and only if $\{\overline{\varphi_n}\}$ is a basis.

3.4 Show that if $\{\varphi_n\}$, $n \in \mathbb{X}$, is a Riesz basis for H then, for each $f \in$ H,

$$\|f\|^2 \asymp \sum_{n \in \mathbb{X}} |\langle f, \varphi_n^* \rangle|^2, \qquad \|f\|^2 \asymp \sum_{n \in \mathbb{X}} |\langle f, \varphi_n \rangle|^2.$$

If the implied bounds in the first relation are A and B, then those in the second are $1/B$ and $1/A$ respectively (Young 1980, p. 36).

3.5 Show that among all sets of coefficients in a frame expansion for a particular Hilbert space member, those of the canonical expansion have the largest l^2 norm.

3.6 Let the members of a sequence (λ_n) belong to a strip of the complex plane which is parallel to the real axis. Then if $\{e^{i\Re e \lambda_n x}\}$ is a frame for $L^2(-\pi w, \pi w)$, so is $\{e^{i\lambda_n x}\}$.

4

FINITE SAMPLING

In this chapter we consider the reconstruction of functions defined on domains with finite Lebesgue measure, from samples taken at finitely many points of the domain. We take our cue from the fact that finite interpolation problems arose first in the historical order of events as outlined in Chapter 1; but our emphasis from now on is going to be on sampling and reconstruction rather than interpolation.

The development of the Lagrange formula, which, in both its finite and its infinite forms applies to functions defined on all of $\mathbb{R}$, will be left to later chapters.

The functions to be represented by finite sampling series will be "polynomials", in the generalized sense that they are truncated expansions in orthogonal sets of various kinds. Because of this they can be said to satisfy an analogue of band-limitation. This is most immediately recognizable in the case of trigonometrical polynomials since they have only finitely many frequency components, frequency being understood in the ordinary sense.

These "polynomials" are the simplest functions of interest in sampling theory. They approximate functions belonging to more general classes, and they provide an amenable research tool in that convergence questions are absent — and sampling in practice is always finite after all!

Brown (1985) has placed the trigonometrical polynomial case in a finite-dimensional rk Hilbert space setting; we find in the following section that these considerations can be made much more general. Finite methods in sampling have often been emphasized by Dodson, Beaty and their co-workers; for example, in Dodson and Silva (1985, p. 84 ff.).

4.1 A general setting for finite sampling

Given a bounded subset E of $\mathbb{R}$, or of $\mathbb{R}^M$, we consider the question of representing any member of a suitable class of functions defined on E by a finite sampling series; that is, in terms of its samples taken at finitely many points distributed in some way over E.

First, suppose that $\{\varphi_n\}$, $n \in \mathbb{N}$, is an orthonormal basis for $L^2(E)$. Then an appropriate function class for finite sampling is the complex Euclidean vector space, or finite-dimensional Hilbert space H_N, of all "polynomials"

$$f(t) = \sum_{n=1}^{N} c_n \varphi_n(t), \tag{4.1.1}$$

for some fixed $N \in \mathbb{N}$, where $\{c_n\} \subset \mathbb{C}$.

If g is another function of this type with coefficients $\{c'_n\}$, the inner product of f and g is

$$\langle f, g \rangle = \sum_{n=1}^{N} c_n \overline{c'_n}.$$

Now it is an immediate consequence of this formula for inner products that H_N has reproducing kernel given by

$$K_N(t, \lambda) = \sum_{i=1}^{N} \varphi_i(t) \overline{\varphi_i(\lambda)}. \tag{4.1.2}$$

This is because

$$f(\lambda) = \sum_{n=1}^{N} c_n \varphi_n(\lambda) = \langle f, K_N(\cdot, \lambda) \rangle,$$

which gives the reproducing equation

These facts follow from the general properties of Hilbert spaces with reproducing kernel given in §3.4, but in this finite-dimensional case it has probably been more instructive to derive them directly.

Now suppose there exists a set of points $\{\lambda_i\} \subset E$, $i = 1, \ldots, N$, such that $\{K_N(t, \lambda_i)\}$ is an orthogonal set in H_N; that is,

$$\langle K_N(\cdot, \lambda_j), K_N(\cdot, \lambda_i) \rangle = K_N(\lambda_i, \lambda_j) \delta_{ij} \qquad (i, j = 1, \ldots, N). \tag{4.1.3}$$

Then $\{K_N(t, \lambda_i)\}$ is a spanning set; in fact, $\{K_N(t, \lambda_i)/\sqrt{K_N(\lambda_i, \lambda_i)}\}$ is an orthonormal basis for H_N in the Hilbert space sense. In this case the expansion for f is the finite sampling series

$$f(t) = \sum_{n=1}^{N} f(\lambda_n) \frac{K_N(t, \lambda_n)}{K_N(\lambda_n, \lambda_n)}, \tag{4.1.4}$$

because its coefficients are given by

$$\left\langle f, \frac{K_N(\cdot, \lambda_n)}{\sqrt{K_N(\lambda_n, \lambda_n)}} \right\rangle = \frac{f(\lambda_n)}{\sqrt{K_N(\lambda_n, \lambda_n)}}.$$

Example 4.1 (Sampling on the Circle) Here the circle $\mathbb{T}$ is taken to mean the set of all real numbers modulo 2π. The set $\{e^{ijt}/\sqrt{2\pi}\}$, $j \in \mathbb{Z}$, is an orthonormal basis for $L^2(\mathbb{T})$, so we take H_N to be the set of trigonometrical polynomials

$$p(t) = \frac{1}{\sqrt{2\pi}} \sum_{j=-N}^{N} c_j e^{ijt}.$$

We shall recover Cauchy's classical formula for trigonometrical interpolation. From (4.1.2) we have the reproducing kernel

$$K_N(t, \lambda) = \frac{1}{2\pi} \sum_{j=n}^{N} e^{ijt} e^{-ij\lambda} = \frac{1}{2\pi} \frac{\sin \frac{1}{2}(2n+1)(t-\lambda)}{\sin \frac{1}{2}(t-\lambda)},$$

by a standard trigonometrical summation (this gives a kernel of the Dirichlet type). Now we can take $t_j = 2\pi j/(2n+1)$, $j = -N, \ldots, N$, and with these specializations (4.1.4) gives Cauchy's formula (1.2.2).

Example 4.2 (Sampling on a Line Segment) We take E to be the interval $[-1, 1]$, and use the simplest classical polynomial basis

$$\varphi_n(t) := (n + \tfrac{1}{2})^{1/2} P_n(t), \qquad P_n(t) = \frac{1}{2^n n!} \frac{d^n}{dt^n} (t^2 - 1)^n, \qquad n \in \mathbb{N}_0,$$

where P_n is the nth Legendre polynomial. It is well known (see for example Higgins 1977, p. 37) that this choice for $\varphi_n(t)$ gives an orthonormal basis for $L^2(-1, 1)$. Hence we take H_N to consist of all functions of the form

$$f(t) = \sum_{n=0}^{N} c_n (n + \tfrac{1}{2})^{1/2} P_n(t), \qquad \{c_n\} \subset \mathbb{C}. \tag{4.1.5}$$

By a special summation formula (Magnus *et al.* 1966, p. 206) we obtain from (4.1.2) the reproducing kernel

$$K_N(t, \lambda) = \sum_{n=1}^{N} (n + \tfrac{1}{2}) P_n(t) P_n(\lambda) = \frac{N+1}{2} \frac{P_{N+1}(\lambda) P_N(t) - P_N(\lambda) P_{N+1}(t)}{t - \lambda}.$$

This is the *Christoffel–Darboux formula* for the Legendre polynomials.[1]

In order to find a set $\{\lambda_n\}$ of $N+1$ sampling points belonging to $[-1, 1]$ such that $K(\lambda_n, \lambda_m) = 0$ we evidently need to solve an equation of the form

$$\frac{P_N(t)}{P_{N+1}(t)} = C,$$

for $N+1$ roots, where C is a constant. In order to do this we need the following facts about the zeros of Legendre polynomials (Magnus *et al.* 1966, p. 206). Every Legendre polynomial $P_N(t)$ has N distinct zeros, all of which lie in the interval $(-1, 1)$; furthermore, the zeros of two successive Legendre polynomials P_N and P_{N+1} are interlaced. Using these facts, a simple consideration of the graphs shows that for any $C > 0$, $P_N(t) = CP_{N+1}(t)$ has $n+1$ distinct roots in $(-1, 1)$.

---

[1] Sometimes called the Christoffel summation formula, as in the book of Sansone (1959, p. 179). This book contains several further interesting summation formulae of the Christoffel type.

Now if $\{\lambda_n\}$ is any such set of roots, we have the sampling representation

$$f(t) = \frac{N+1}{2} \sum_{n=0}^{N} f(\lambda_n) \frac{P_{N+1}(\lambda_n)P_N(t) - P_N(\lambda_n)P_{N+1}(t)}{t - \lambda_n},$$

and it holds for every f of the form (4.1.5).

$$* \quad * \quad *$$

Point sets $\{\lambda_i\}$ with the properties outlined above may or may not exist. However, it may be possible to find $\{\lambda_i\}$ for which $\{K_N(t, \lambda_i)\}$ is a spanning set for H_N, even though it is not orthogonal. We can still formulate a sampling theory, although it is not quite so convenient as in the orthogonal case. It arises because $\{K_N(t, \lambda_i)\}$ will be complete in H_N in the Hilbert space sense, and because H_N is finite-dimensional it *must* then possess a unique biorthonormal set $\{\Psi_j\}$, $j = 1, \ldots, N$, say; that is, a set such that

$$\langle K_N(\cdot, \lambda_i), \Psi_j \rangle = \delta_{ij}.$$

The expansion for $f \in H_N$ in the set $\{\Psi_j\}$ is of the form

$$f(t) = \sum_{n=1}^{N} b_n \Psi_n(t)$$

in which $b_n = \langle f, K_N(\cdot, \lambda_n) \rangle = f(\lambda_n)$, so we have the sampling representation

$$f(t) = \sum_{n=1}^{N} f(\lambda_n) \Psi_n(t).$$

A useful criterion by which one can check for completeness numerically is

$$\Delta := \det(\varphi_n(\lambda_j)) \neq 0. \tag{4.1.6}$$

To obtain this criterion we notice that the condition for completeness is

$$\langle f, K_N(\cdot, \lambda_i) \rangle = f(\lambda_i) = 0 \implies f \equiv 0, \qquad i = 1, \ldots, N.$$

But from (4.1.1) this can be written

$$\sum_{n=1}^{N} c_n \varphi_n(\lambda_i) = 0 \implies f \equiv 0, \qquad i = 1, \ldots, N.$$

Now if $\Delta \neq 0$, there is only the trivial solution $c_i = 0$, $i = 1, \ldots, N$, to this set of equations; hence f is null.

To calculate the Ψ_j's, first put

$$\Psi_j(t) = \sum_{k=1}^{N} c_k^{(j)} \varphi_k(t).$$

Now

$$\langle \Psi_j, K_N(\cdot, \lambda_i) \rangle = \delta_{ij} = \Psi_j(\lambda_i) = \sum_{n=1}^{N} c_n^{(j)} \varphi_n(\lambda_i).$$

Hence we need to solve the set of equations

$$\sum_{n=1}^{N} c_n^{(j)} \varphi_n(\lambda_i) = \delta_{ij} \tag{4.1.7}$$

for the coefficients $\{c_n^{(j)}\}$ of Ψ_j for $j = 1, \ldots, N$. The criteria (4.1.7) and (4.1.6) will be needed in Problem 4.2.

4.2 Sampling on the sphere

Phenomena in which functions have a spherical surface as their domain of definition arise naturally in the physical sciences; for example, in the study of earthquakes, the hydrogen atom and the solar corona. A potential application is to medical imaging, particularly of the brain, where the problem is the optimal placement of electroencephalograph electrodes on a human patient's head.[2]

An appropriate Fourier analysis on the sphere was developed many years ago; Laplace and Legendre introduced expansions of functions in spherical harmonics in the 1780s in order to study gravitational theory. It therefore seems to be a natural question to ask for a theory for sampling at finitely many points distributed over a spherical surface.

Let S_2 denote the surface of the unit sphere in $\mathbb{R}^3$, and let T be a point on S_2 with the usual co-ordinates φ of polar co-latitude and θ of longitude. For $L^2(S_2)$ we shall use the classical orthonormal basis of surface spherical harmonics

$$\Phi_n(\varphi) := \frac{1}{\sqrt{4\pi}} \sqrt{2n+1} P_n(\cos\varphi)$$

$$\Phi_{nm}^{(c)} := \frac{1}{\sqrt{4\pi}} \sqrt{\frac{2(2n+1)(n-m)!}{(n+m)!}} \cos m\theta P_n^m(\cos\varphi)$$

$$\Phi_{nm}^{(s)} := \frac{1}{\sqrt{4\pi}} \sqrt{\frac{2(2n+1)(n-m)!}{(n+m)!}} \sin m\theta P_n^m(\cos\varphi),$$

---

[2]This was kindly explained to the author by Professor Victor Pollak at a conference at Aachen, Germany, in 1987.

for all $n \in \mathbb{N}$ and $m = 1, \ldots, n$, where P_n is the Legendre polynomial defined in Example 4.2, and P_n^m is the associated Legendre function of the first kind defined by

$$P_n^m(u) := (1 - u^2)^{m/2} \frac{d^m}{du^m} P_n(u).$$

The expansion for $f \in L^2(S_2)$ is

$$f(\varphi, \theta) = a_0 \Phi_0 + \sum_{n=1}^{\infty} \left\{ a_n \Phi_n(\varphi) + \sum_{m=1}^{n} a_{nm} \Phi_{nm}^{(c)}(\varphi, \theta) + b_{nm} \Phi_{nm}^{(s)}(\varphi, \theta) \right\}. \tag{4.2.1}$$

This expansion will be instrumental in identifying the space H_N for our sampling theory. We must find a suitable truncation of it, and of the many ways of doing this the termination of n at an integer κ will be a particularly convenient one. It is easy to check that the resulting finite series will then contain $(\kappa + 1)^2$ terms. So with $N = (\kappa + 1)^2$, the space H_N will consist of linear combinations of the first N surface spherical harmonics, where N is a squared integer.

Let $U : (\varphi', \theta')$ be another point of S_2. The reproducing kernel K_N is, from (4.1.2), given by

$$K_N(T, U) = \sum_{n=0}^{\kappa} \left\{ \Phi_n(\varphi) \Phi_n(\varphi') + \sum_{m=1}^{n} \Phi_{nm}^{(c)}(\varphi, \theta) \Phi_{nm}^{(c)}(\varphi', \theta') + \Phi_{nm}^{(s)}(\varphi, \theta) \Phi_{nm}^{(s)}(\varphi', \theta') \right\}.$$

We are now going to benefit from the power of Legendre's addition formula (Whittaker and Watson 1962, p. 395). It says that the summand in braces above is just $P_n(x)$ if x is defined by

$$x = \cos \varphi \cos \varphi' + \sin \varphi \sin \varphi' \cos(\theta - \theta').$$

We then have

$$K_N(T, U) = \frac{1}{4\pi} \sum_{n=0}^{\kappa} (2n + 1) P_n(x) = \frac{1}{4\pi} \sum_{n=0}^{\kappa} \left[P_{n+1}'(x) - P_{n-1}'(x) \right]$$

$$= \frac{1}{4\pi} \left[P_\kappa'(x) + P_{\kappa+1}'(x) \right],$$

the second equality arising from a recurrence relation for P_n, and the last from collapsing the summation.

Let us now describe very briefly the special case $N = 4$; the reader is invited to supply details (or refer to Kempski 1995, §4.2). First, one calculates

$$K_4(T,U) = \frac{1}{4\pi}(3x+1).$$

Then, on taking one point to be the north pole, some testing reveals that four points on S_2 for which the orthogonality relation (4.1.3) holds are

$$T_1 : (0,0), \; T_2 : \left(\arccos\left(-\tfrac{1}{3}\right),0\right), \; T_3 : \left(\arccos\left(-\tfrac{1}{3}\right),\tfrac{2\pi}{3}\right), \; T_4 : \left(\arccos\left(-\tfrac{1}{3}\right),\tfrac{4\pi}{3}\right).$$

They are the vertices of a regular tetrahedron inscribed in S_2. The sampling series (4.1.4) for $f \in H_4$ is

$$\begin{aligned}
f(T) = \frac{1}{4\pi}\Big\{ & f(T_1)\,(3\cos\varphi + 1) \\
& + f(T_2)\,(-\cos\varphi + 2\sqrt{2}\sin\varphi\cos\theta + 1) \\
& + f(T_3)\,(-\cos\varphi - \sqrt{2}\sin\varphi\cos\theta + \sqrt{2}\sqrt{3}\sin\varphi\sin\theta + 1) \\
& + f(T_4)\,(-\cos\varphi - \sqrt{2}\sin\varphi\cos\theta - \sqrt{2}\sqrt{3}\sin\varphi\sin\theta + 1) \Big\}.
\end{aligned}$$

It should be mentioned that caution has been urged in the use of spherical harmonics for sampling purposes, by Freeden (1984), because of "their strongly oscillating properties"; spherical spline interpolation being the recommended alternative. The present methods were introduced by Higgins (1987a), and much computational detail supplied by Kempski (1995, §4.2).

PROBLEMS

4.1 Show that orthogonality, which worked for the case $N = 4$ in the text, fails in the case $N = 9$.

4.2 Find a sampling theorem in the case of six points that are "fairly distributed" over S_2. Hint: consider the vertices $O_1, \ldots, O_6$ (one of which can be taken as the north pole) of an inscribed regular octahedron. Identify ten six-term truncations of (4.2.1) by taking the first four terms together with two chosen from the term corresponding to $n = 2$. After identifying a suitable kernel, $K_6(S,T)$, say, you will need to verify the completeness criterion (4.1.6), and then calculate the biorthogonal expasion functions from (4.1.7).

Then if $f(T)$ denotes any of these six-term truncations, the sampling representation is

$$f(T) = f(O_1) \frac{1}{2} \cos \varphi (\cos \varphi + 1)$$

$$+ f(O_2) \frac{1}{4} (1 + 2 \sin \varphi \cos \theta - \cos^2 \varphi + \sin^2 \varphi \cos 2\theta)$$

$$+ f(O_3) \frac{1}{4} (1 + 2 \sin \varphi \cos \theta - \cos^2 \varphi - \sin^2 \varphi \cos 2\theta)$$

$$+ f(O_4) \frac{1}{4} (1 - 2 \sin \varphi \cos \theta - \cos^2 \varphi + \sin^2 \varphi \cos 2\theta)$$

$$+ f(O_5) \frac{1}{4} (1 - 2 \sin \varphi \cos \theta - \cos^2 \varphi - \sin^2 \varphi \cos 2\theta)$$

$$+ f(O_6) \frac{1}{2} \cos \varphi (\cos \varphi - 1).$$

Solve the same problem in the case of twenty points.
Comment on the notion of "fair distribution" of points on a spherical surface.

For more details on these problems, see Kempski (1995, §4.2).

5

FROM FINITE TO INFINITE SAMPLING SERIES

An interesting question, which has come up from time to time in the research literature, is whether sampling theorems involving infinitely many sampling points can be derived as limiting cases of finite sampling theorems. Indeed, the cardinal series arises as a formal limiting case of the finite Lagrange formula (the right-hand side of (1.2.1)), as we shall mention in a moment; but a fully rigorous treatment of this limiting procedure was not given until very recently. The present chapter is devoted entirely to the discussion of this problem by Hinsen and Klösters (1993).

First, let us modify the Lagrange formula by allowing a "two way" sequence of sampling points, and write it in terms of the samples of a given function f. For convenience the sample points are "normalized" so that $\lambda_0 = 0$. The result is

$$H_M(t) \left[\frac{f(0)}{t} + \sum_{j=1}^{M} \left\{ \frac{f(\lambda_j)}{H'_M(\lambda_j)(t - \lambda_j)} + \frac{f(\lambda_{-j})}{H'_M(\lambda_{-j})(t - \lambda_{-j})} \right\} \right],$$

where

$$H_M(t) = t \prod_{j=1}^{M} \left(1 - \frac{t}{\lambda_j} \right) \left(1 - \frac{t}{\lambda_{-j}} \right).$$

In this form of Lagrange's formula let us put $\lambda_j = j$, $j \in \mathbb{Z}$, and pass to the *formal* limit as $M \to \infty$. Because of Euler's product representation

$$\frac{\sin \pi t}{\pi t} = \prod_{k=1}^{\infty} \left(1 - \frac{t^2}{k^2} \right), \tag{5.0.1}$$

we obtain the cardinal series in the form

$$\frac{\sin \pi t}{\pi} \left\{ \frac{f(0)}{t} + \sum_{n=1}^{\infty} (-1)^n \left[\frac{f(n)}{t - n} + \frac{f(-n)}{t + n} \right] \right\}.$$

The left-hand side of (1.2.1) can be given the same treatment that we have just given to the right-hand side. In this way one can obtain formally the Newton–Gauss formula of interpolation (a discussion of the interrelationships between these series can be found in J.M. Whittaker (1935, Ch. IV)).

5.1   The change to infinite sampling series

We now turn to the method of Hinsen and Klösters. They report that a start on justifying the formal limiting proceedure mentioned above was made by Ferrar in 1925, and he seems to have been the first to introduce de la Valleé Poussin summability into the theory. Ferrar showed that

$$\lim_{r \to \infty} \frac{\omega_r(t)}{\omega_r'(n)(t-n)} = \operatorname{sinc}(t-n),$$

where

$$\omega_r(t) := t \prod_{k=1}^{r} (t^2 - k^2). \tag{5.1.1}$$

However, this only settles part of the problem. The proof of Hinsen and Klösters continues to use an interesting technique of approximation theory involving de la Valleé Poussin's convergence factors

$$\theta_{r,n} := \frac{(r!)^2}{(r-n)!(r+n)!}, \qquad r \geq |n|.$$

Two properties of these factors will be needed. The first is

$$|\theta_{r,n}| \leq 1. \tag{5.1.2}$$

This follows from the fact that $\theta_{r,0} = 1$, $\theta_{r,-n} = \theta_{r,n}$ and

$$\theta_{r,n} = \frac{r!/(r-n)!}{(r+n)!/r!} = \prod_{j=0}^{n-1} \frac{r-j}{r+n-j} \leq 1, \qquad 0 < n \leq r.$$

The second property that we shall need is

$$\lim_{r \to \infty} \theta_{r,n} = 1, \qquad n \in \mathbb{Z}. \tag{5.1.3}$$

This is easily proved using Stirling's formula (Magnus *et al.* 1966, p. 12).

Let $C[a, b]$ denote the Banach space of functions which are continuous on the closed bounded interval $[a, b]$, with norm $\|\varphi\|_C := \max_{t \in [a,b]} |\varphi|$. Let

$$(\mathcal{L}_r f)(t) := \sum_{|n| \leq r} f(n) \frac{\omega_r(t)}{\omega_r'(n)(t-n)}, \qquad t \in \mathbb{R},$$

where ω_r is defined in (5.1.1); let

$$(\mathcal{C}f)(t) := \sum_{n \in \mathbb{Z}} f(n) \operatorname{sinc}(t-n), \quad \text{and} \quad (\mathcal{C}_r f)(t) := \sum_{|n| \leq r} f(n) \operatorname{sinc}(t-n).$$

The main result of this chapter is:

Theorem 5.1 *Let f be defined on $\mathbb{Z}$, and let $(f(n)) \in l^p$, $1 \leq p < \infty$. Then*

$$\lim_{r \to \infty} \|\mathcal{L}_r f - \mathcal{C}f\|_C \to 0 \qquad (5.1.4)$$

where $C[a, b]$ is defined above.

The proof depends on the following:

Lemma 5.2 *Let $1 \leq p < \infty$ and let q be the usual conjugate index given by $1/p + 1/q = 1$.*

(a) If (w_n), $n \in \mathbb{Z}$, is a sequence belonging to l^p and $(\psi_n(t))$, $n \in \mathbb{Z}$, is a sequence of continuous functions defined on $[a, b]$ such that

$$\begin{cases} \sup_{t \in [a,b]} \left(\sum_{n \in \mathbb{Z}} |\psi_n(t)|^q \right)^{1/q} = C_{\psi,q} < \infty, & 1 < q < \infty; \\[2em] \sup_{t \in [a,b]} \sup_{n \in \mathbb{Z}} |\psi_n(t)| = C_{\psi,\infty} < \infty, & q = \infty; \end{cases} \qquad (5.1.5)$$

then the series

$$\sum_{n \in \mathbb{Z}} w_n \psi_n(t) \qquad (5.1.6)$$

converges absolutely and uniformly over $[a, b]$, the sum of (5.1.6) being a continuous function.

(b) If, in addition to the hypotheses of part (a), we suppose that for $r \in \mathbb{N}$ and $n \in \mathbb{Z}$, $|n| \leq r$, there exist continuous functions $\psi_{r,n}$ such that

$$\lim_{r \to \infty} \|\psi_{r,n} - \psi_n\|_C = 0, \qquad n \in \mathbb{Z}, \qquad (5.1.7)$$

and further, there exist constants $\alpha > 1/q$ and $C_{\alpha,\infty} > 0$ for which

$$\|\psi_{r,n}\|_C \leq C_{\alpha,\infty} |n|^{-\alpha} \qquad (5.1.8)$$

whenever $|n| > n_0$ for some $n_0 \in \mathbb{N}_0$. Then

$$\lim_{r \to \infty} \left\| \sum_{|n| \leq r} w_n \theta_{r,n} \psi_{r,n} - \sum_{|n| \leq r} w_n \psi_n \right\|_C = 0.$$

Proof (a) Whenever $s < r$ we have, by Hölder's inequality and (5.1.5),

$$\left| \sum_{s \leq |n| \leq r} w_n \psi_n(t) \right| \leq \sum_{s \leq |n| \leq r} |w_n| |\psi_n(t)|$$

$$\leq \begin{cases} \displaystyle\sum_{s \leq |n| \leq r} |w_n| \max_{s \leq |n| \leq r} |\psi_n(t)|, & p = 1 \\[3ex] \displaystyle\left(\sum_{s \leq |n| \leq r} |w_n|^p \right)^{1/p} \left(\sum_{s \leq |n| \leq r} |\psi_n|^q \right)^{1/q}, & 1 < p < \infty \end{cases}$$

$$\leq C_{\psi,q} \left(\sum_{s \leq |n| \leq r} |w_n|^p \right)^{1/p}, \qquad t \in [a, b].$$

The proof of part (a) is now completed by applying Cauchy's convergence criterion; indeed, the series $\sum_{s \leq |n| \leq r} w_n \psi_n(t)$ and $\sum_{s \leq |n| \leq r} |w_n| |\psi_n(t)|$ are uniformly convergent and hence their sums are continuous functions.

(b) Let $\epsilon > 0$. Since $(w_n) \in l^p$ there exists a positive integer $N > n_0$ such that

$$\left(\sum_{|n| > N} |w_n|^p \right)^{1/p} < \epsilon. \tag{5.1.9}$$

Now from (5.1.5), (5.1.7) and (5.1.8) we find that, for any fixed N and arbitrary $r > N$,

$$\begin{cases} \left\| \left(\displaystyle\sum_{N < |n| \leq r} |\psi_n|^q \right)^{1/q} \right\| \leq C_{\psi,q}, & 1 < q < \infty; \\[4ex] \left\| \max_{N < |n| \leq r} |\psi_n| \right\| \leq C_{\psi,q}, & q = \infty. \end{cases} \tag{5.1.10}$$

Furthermore, there is a constant $C_{\alpha,q}$, say, depending only on α and q, such that

$$\left\{ \begin{array}{l} \left\|\left(\sum_{N<|n|\leq r}|\psi_{r,n}(t)|^q\right)^{1/q}\right\| \leq C_\alpha\left(\sum_{N<|n|\leq r}|n|^{-\alpha q}\right)^{1/q} \\[2em] \left\|\max_{N<|n|\leq r}|\psi_{r,n}(t)|\right\| \leq C_{\alpha,\infty}(N+1)^{-\alpha} \end{array}\right\} \leq C_{\alpha,q}, \qquad (5.1.11)$$

because the series $\sum_{n\in\mathbb{N}}n^{-\alpha q}$ is, by hypothesis, convergent. Again, there is a positive constant M_N, say, such that

$$\|\psi_{r,n}\| \leq M_N, \qquad |n| \leq N. \qquad (5.1.12)$$

Now (5.1.7) and (5.1.3) show that there exists a number $R_{N,\epsilon} \geq N$ such that, for $|n| \leq N$ and $r \geq R_{N,\epsilon}$ we have

$$\|\psi_{r,n} - \psi_n\| \leq \frac{\epsilon}{1 + \sum_{|n|\leq N}|w_n|} \qquad (5.1.13)$$

and

$$|\theta_{r,n} - 1| \leq \frac{\epsilon}{M_N(1 + \sum_{|n|\leq N}|w_n|)}. \qquad (5.1.14)$$

We can now complete the proof by using Minkowski's inequality, Theorem 2.4, Hölder's inequality, Theorem 2.3, (5.1.2) and the estimates (5.1.9)–(5.1.14) as follows. For $r > R_{N,\epsilon}$,

$$\left\|\sum_{|n|\leq r}w_n\theta_{r,n}\psi_{r,n} - \sum_{|n|\leq r}w_n\psi_n\right\|$$

$$\leq \sum_{|n|\leq N}|w_n|\,\|\theta_{r,n}\psi_{r,n} - \psi_n\| + \sum_{N<|n|\leq r}|w_n|(|\theta_{r,n}|\,|\psi_{r,n}| + |\psi_n|)$$

$$\leq \sum_{|n|\leq N}|w_n|(|\theta_{r,n} - 1|\,\|\psi_{r,n}\| + \|\psi_{r,n} - \psi_n\|)$$

$$+ \left\|\sum_{N<|n|\leq r}|w_n|\,|\psi_{r,n}|\right\| + \left\|\sum_{N<|n|\leq r}|w_n|\,|\psi_n|\right\|$$

$$\leq \sum_{|n|\leq N}|w_n|\left\{\frac{\epsilon M_N}{M_N(1 + \sum_{|n|\leq N}|w_n|)} + \frac{\epsilon}{(1 + \sum_{|n|\leq N}|w_n|)}\right\}$$

$$+ (C_{\alpha,q} + C_{\psi,q})\left(\sum_{N<|n|\leq r}|w_n|^p\right)^{1/p}$$

$$\leq \epsilon(2 + C_{\alpha,q} + C_{\psi,q}).$$

5.2 The theorem of Hinsen and Klösters

We can now prove Theorem 5.1. We take $\psi_n(t)$ of the previous lemma to be $\mathrm{sinc}(t-n)$, $n \in \mathbb{Z}$. It will be proved in Lemma 11.2 that $\|(\mathrm{sinc}(t-n))\|_{lq}^q < p$, $1 < p < \infty$, $t \in \mathbb{R}$, and because $\max_{t \in \mathbb{R}} \mathrm{sinc}\, t = 1$ the estimates in (5.1.5) hold for all $t \in [a,b]$.

Next we choose $\psi_{r,n}(t) := (-1)^n (t-n)^{-1} t \prod_{k=1}^r (1 - t^2/k^2)$, $r \geq |n|$. Because the convergence in the product representation (5.0.1) is uniform over $[a,b]$ there is a constant, $M_{a,b}$ say, such that

$$\left\| t \prod_{k=1}^r \left(1 - \frac{t^2}{k^2}\right) \right\| \leq M_{a,b}, \qquad r \in \mathbb{N}. \tag{5.2.1}$$

It is clear that as $r \to \infty$, $\psi_{r,n}(t)$ tends to $\psi_n(t)$ pointwise on $\mathbb{R}$; in fact the convergence is uniform over $[a,b]$ because

$$\lim_{r \to \infty} \psi_{r,n}(t) = \begin{cases} t \displaystyle\prod_{k=1}^{|n|-1} \left(1 - \frac{t^2}{k^2}\right) \frac{(-1)^{n-1}}{n} \left(1 + \frac{t}{n}\right) \prod_{k=|n|+1}^{\infty} \left(1 - \frac{t^2}{k^2}\right), & |n| \in \mathbb{N}, \\[4mm] \displaystyle\prod_{k=1}^{\infty} \left(1 - \frac{t^2}{k^2}\right), & n = 0, \end{cases}$$

and we have, for $t \in [a,b]$, $|t^2/k^2| \leq \max\{|a|^2, |b|^2\}/k^2$, and of course $\sum_{k \in \mathbb{N}} k^{-2} < \infty$, so the uniformity follows by the standard criterion. Hence (5.1.7) holds.

Suppose that $r \geq |n| \geq n_0$, where n_0 is larger by 1 than the smallest integer exceeding or equal to $\max\{|a|, |b|\}$. Then by (5.2.1) we have

$$\|\psi_{r,n}\| \leq \max_{t \in [a,b]} \frac{M_{a,b}}{|t-n|} \leq \frac{M_{a,b}}{|n| - \max\{|a|, |b|\}} \leq M_{a,b} n_0 |n|^{-1},$$

so that (5.1.7) holds with $\alpha = 1$ (which exceeds $1/q$ when $1 < q < \infty$), and $C_{\alpha, \infty} = M_{a,b} n_0$. From (5.1.1) it follows that

$$\omega'_r(n) = \prod_{\substack{k=-r, \\ k \neq n}}^{r} (n-k) = (-1)^{r+n} (r-n)! (r+n)!, \qquad |n| \leq r.$$

This implies that

$$\sum_{n=-r}^{r} f(n) \theta_{r,n} \psi_{r,n} = \sum_{n=-r}^{r} f(n) \frac{r!}{(r-n)!(r+n)!} \frac{1}{(-1)^n (t-n)} t \prod_{k=1}^{r} \left(1 - \frac{t^2}{k^2}\right)$$

$$= \sum_{n=-r}^{r} f(n) \frac{\omega_r(t)}{\omega'_r(n)(t-n)} = (\mathcal{L}_r f)(t).$$

Finally, if w_n is chosen to be $f(n)$, $n \in \mathbb{Z}$, Lemma 5.2 part (b) shows that $\lim_{r\to\infty} \|\mathcal{L}_r f - C_r f\| = 0$, part (a) shows that $\lim_{r\to\infty} \|C_r f - Cf\| = 0$ and the required result follows from

$$\|\mathcal{L}_r f - Cf\| \le \|\mathcal{L}_r f - C_r f\| + \|C_r f - Cf\|.$$

It should be emphasized that the foregoing theorem is not, in itself, a proof of a sampling theorem. In fact it holds under much weaker hypotheses than those of theorems governing the representation of functions in cardinal series. It is important to link the ideas together, however, and in the following corollary it is shown that when f belongs to B_π^p (the *Bernstein space* defined in Definition 6.5) the limits of the Lagrange operator $\mathcal{L}$ and of the sampling series operator C acting on f give the same result, namely a representation of f. Here the reader is asked to look forward to Chapter 6 for Theorems 6.8 and 6.13, which are needed for the proof.

Corollary 5.3 *Let* $f \in B_\pi^p$, $1 \le p < \infty$. *Then*

$$\lim_{r\to\infty} (\mathcal{L}_r f)(t) = (Cf)(t) = f(t)$$

uniformly on any bounded interval $[a, b]$.

Proof After the previous theorem we need only check that if $f \in B_\pi^p$ then $(f(n)) \in l^p$ to establish the first equality. But this follows from Nikol'skii's inequality, Theorem 6.8. The second equality will be proved in Theorem 6.13.

A more general form of Theorem 5.1, in which the sample points are not uniformly spaced, has also been given by Hinsen and Klösters (1993).

6

SAMPLING FOR BERNSTEIN AND PALEY–WIENER SPACES

The Bernstein classes B_σ^p and the Paley–Wiener classes PW_σ^p, $\sigma > 0$, $1 \le p \le \infty$, are particularly important for our study of the cardinal series and some of its generalizations. Before studying the basic sampling theories for these classes we need preliminary definitions about entire functions of exponential type, and some principles for working with Bernstein classes. This occupies §6.1; no proofs are given here, but references to standard sources are included. The books of Boas (1954), Nikol'skiĭ (1975) and Young (1980) provide excellent sources for this and much further relevant material.

After this the "ordinary" sampling theory for Bernstein and Paley–Wiener functions is developed. Here, ordinary is taken to mean that sampling is *regular* in that the sample points are equidistantly spaced, the functions to be sampled have one-dimensional domains and they are band-limited in the conventional or the distributional sense to band-regions that consist of a single interval centred at the origin. In later chapters these conditions will be relaxed.

6.1 Bernstein spaces

Definition 6.1 *The function $f(z)$ is said to be entire if it is holomorphic throughout $\mathbb{C}$.*

Definition 6.2 *An entire function is said to be of exponential type if there exist positive constants A and B such that*

$$|f(z)| \le Ae^{B|z|}, \qquad z \in \mathbb{C}.$$

Definition 6.3 *Associated with such a function is a number τ called the exponential type of f, defined by*

$$\tau = \limsup_{r \to \infty} \frac{\log M(r)}{r},$$

where $M(r)$ denotes the maximum modulus of f on the circle $|z| = r$.

There are three conditions which are often used when working with functions of exponential type. The first is that the exponential type of such a function f is a number which does not exceed τ if, given $\epsilon > 0$, there is a constant A_ϵ such that

$$|f(z)| \le A_\epsilon e^{(\tau+\epsilon)|z|}. \tag{6.1.1}$$

One says that "f is of exponential type at most τ".

A slightly stronger condition is

$$|f(z)| \le A e^{\tau|z|}, \qquad z \in \mathbb{C}. \tag{6.1.2}$$

If f satisfies (6.1.2) then of course it satisfies (6.1.1), and so it is of exponential type at most τ.

A condition which is slightly stronger still is

$$e^{-\tau|z|} |f(z)| = o(1) \qquad \text{for large } |z|. \tag{6.1.3}$$

If f satisfies (6.1.3) then it satisfies (6.1.2), and so in turn it satisfies (6.1.1).

Definition 6.4 *By E_σ we denote the class whose members are entire functions in the complex plane, and are of exponential type at most σ.*

Definition 6.5 *The Bernstein class B_σ^p consists of those functions which belong to E_σ and whose restriction to $\mathbb{R}$ belongs to $L^p(\mathbb{R})$.*

The class B_σ^p is a Banach space if one takes the norm to be that of $L^p(\mathbb{R})$ when $p < \infty$, and if one takes the sup norm when $p = \infty$ (Young 1980, pp. 84 and 99). Therefore from now on we shall speak of *Bernstein spaces*.

Lemma 6.6 *We have $B_\sigma^1 \subset B_\sigma^p \subset B_\sigma^r \subset B_\sigma^\infty$, $1 \le p \le r \le \infty$, and $f_1 \in B_\sigma^p$, $f_2 \in B_\sigma^q$ imply $f_1 f_2 \in B_{2\sigma}^1$.*

Of the various inequalities carrying the name Bernstein, the following is most relevant for us.

Theorem 6.7 (Bernstein's Inequality) *For every $f \in B_\sigma^p$, $r \in \mathbb{N}$ and $p \ge 1$, we have*

$$\|f^{(r)}\|_p \le \sigma^r \|f\|_p.$$

Theorem 6.8 (Nikol'skiĭ's Inequality) *Let $1 \le p \le \infty$. Then, for every $f \in B_\sigma^p$ and $h > 0$,*

$$\|f\|_p \le \sup_{u \in \mathbb{R}} \left\{ h \sum_{n \in \mathbb{Z}} |f(u - hn)|^p \right\}^{1/p} \le (1 + h\sigma) \|f\|_p$$

(Nikol'skiĭ 1975, p. 124).

Theorem 6.9 (Compactness Principle for Bernstein Spaces) *Let $1 \le p \le \infty$, and let (F_n), $n \in \mathbb{N}$, be a sequence of members of B_σ^p. Then, given a compact subset C of $\mathbb{C}$, there exists a subsequence (F_{n_k}) of (F_n) that converges uniformly on C.* (Nikol'skiĭ 1975, p. 127).

Our next inequality is quite powerful in that it allows for irregularly distributed points.

Theorem 6.10 (The Plancherel–Pólya Inequality) *Let $f \in B_\sigma^p$, $p > 0$, and let $\Lambda = (\lambda_n)$, $n \in \mathbb{Z}$, be a real increasing sequence such that $\lambda_{n+1} - \lambda_n \geq 2\delta$. Then*

$$\sum_{n \in \mathbb{Z}} |f(\lambda_n)|^p \leq \frac{2e^{p\sigma\delta}}{\pi\delta} \|f\|_p^p.$$

(This inequality can be found in Boas (1954, p. 101)).

It should be noted that the constant on the right-hand side of this inequality is independent of f, but does depend on p, σ and the nature of (λ_n).

Further interesting material about Bernstein spaces is reviewed in Butzer *et al.* (1988).

6.2   Convolution and the cardinal series

The first task that we shall ask of Poisson's summation formula is to help in proving the following important theorem for functions belonging to Bernstein spaces. It will be convenient to take $\sigma = \pi w$.

Theorem 6.11 *Let $f_1 \in B_\sigma^p$ and $f_2 \in B_\sigma^q$, where $1 \leq p \leq \infty$ and q is the conjugate index. Then*

$$\sum_{n \in \mathbb{Z}} f_1\left(\frac{n}{w}\right) f_2\left(\tau - \frac{n}{w}\right) = w \int_{\mathbb{R}} f_1(u) f_2(\tau - u)\, du, \qquad \tau \in \mathbb{R}. \tag{6.2.1}$$

Proof First let $g \in B_{2\sigma}^1$. By the L^1 version of the Paley–Wiener Theorem (Theorem 7.4) we have

$$g(t) = \int_{-2\pi w}^{2\pi w} e^{itu} \varphi(u)\, du.$$

This means that φ is integrable and L^1 Fourier inversion (Theorem 2.11) holds. Hence, φ can be identified with $g^\wedge$ (everywhere, by continuity). We have supp $g^\wedge \subseteq [-2\pi w, 2\pi w]$ and $g^\wedge(\pm 2\pi w) = 0$, so that when Poisson's summation formula (2.3.1) is applied to g it reduces to

$$\sum_{n \in \mathbb{Z}} g\left(\frac{1}{w}(n + t)\right) \sim w\sqrt{(2\pi)} g^\wedge(0). \tag{6.2.2}$$

The series on the left is a sum of analytic functions, and from Nikol'skiĭ's inequality it is absolutely convergent in any closed real interval I containing $t = 0$.

Let us take its sequence of partial sums to be the sequence (F_n) of the compactness principle, Theorem 6.9. Because the series converges, every subsequence of (F_n) converges to the same sum, and we are guaranteed that at least one such subsequence converges uniformly on I.

Since the sum is also periodic (with period 1), we now find that it satisfies sufficient conditions for pointwise representation in Fourier series. Therefore in (6.2.2) we can replace "$\sim$" with "$=$", and putting $t = 0$ we can write

$$\sum_{n \in \mathbb{Z}} g\left(\frac{n}{w}\right) = w\sqrt{2\pi}g^{\wedge}(0) = w \int_{\mathbb{R}} g(u)\, du. \qquad (6.2.3)$$

Here we have added, on the right-hand side, the L^1 definition of the Fourier transform. By the second part of Lemma 6.6 we can take $g(u) = f_1(u)f_2(\tau - u)$ in (6.2.3) for every $\tau \in \mathbb{R}$, and this gives (6.2.1).

Corollary 6.12 *In (6.2.1), f_1 and f_2 on the left-hand side can have their arguments interchanged.*

Proof Since the arguments of f_1 and f_2 can be interchanged on the right-hand side of (6.2.1) by a simple change of variable, they can be interchanged on the left.

Theorem 6.11 (and its corollary) tell us that convolution of Bernstein functions always has an equivalent formulation as a discrete convolution. This "discretization of convolution" has important consequences in approximation theory (Butzer *et al.* 1988, p. 10 ff.); here we are going to use the idea to derive a rather general theorem on the representation of functions in sampling series. But before leaving these remarks about discretization, it is interesting to notice that (6.2.1) has the form of an "integral analogue of a series"; indeed, the right-hand side can be rewritten as an exact integral analogue of the left-hand side, as in the more primitive form, derived from (6.2.3),

$$\sum_{n \in \mathbb{Z}} g\left(\frac{n}{w}\right) = \int_{\mathbb{R}} g\left(\frac{u}{w}\right)\, du.$$

In this form we find, furthermore, an exact quadrature formula of a particularly striking kind.

We come now to a general sampling theorem for Bernstein spaces.

Theorem 6.13 *For every $f \in B^p_{\pi w}$, $1 \leq p < \infty$, we have the cardinal series expansion*

$$f(t) = \sum_{n \in \mathbb{Z}} f\left(\frac{n}{w}\right) \text{sinc}(wt - n). \qquad (6.2.4)$$

The series converges absolutely for $t \in \mathbb{C}$, and uniformly on compact subsets.

Proof In the previous theorem we can take $f_1(t) = f(t)$, and $f_2(t) = \text{sinc}\, wt$; then the corollary gives

$$\sum_{n \in \mathbb{Z}} f\left(\frac{n}{w}\right) \text{sinc}(wt - n) = \sum_{n \in \mathbb{Z}} f\left(t - \frac{n}{w}\right) \text{sinc}(n) = f(t).$$

Now let us apply Hölder's inequality and obtain

$$\sum_{n\in\mathbb{Z}}\left|f\left(\frac{n}{w}\right)\operatorname{sinc}(wt-n)\right| \le \left\{\sum_{n\in\mathbb{Z}}\left|f\left(\frac{n}{w}\right)\right|^p\right\}^{1/p}\left\{\sum_{n\in\mathbb{Z}}|\operatorname{sinc}(wt-n)|^q\right\}^{1/q}.$$

The first sum on the right converges, either by Theorem 6.8 or by Theorem 6.10. We now need to show that the second sum converges uniformly on every compact subset of $\mathbb{C}$.

The reader may care to prove this directly; however, looking forward to Lemma 11.2, a remarkable estimate for this sum due to Splettstößer *et al.* will be proved. In fact the sum is shown to be bounded by $p^{1/q}$, and this is more than adequate for the present task. This completes the proof.

A sampling series representing derivatives of band-limited functions in terms of function samples can now be deduced easily from the ordinary series.

Theorem 6.14 *For every $f \in B_{\pi w}^p$, $1 \le p < \infty$, and $r \in \mathbb{N}$, we have*

$$f^{(r)}(t) = \sum_{n\in\mathbb{Z}} f\left(\frac{n}{w}\right)\left(\frac{d}{dt}\right)^r \operatorname{sinc}(wt-n). \qquad (6.2.5)$$

The series converges absolutely for $t \in \mathbb{C}$, and uniformly on compact subsets.

Proof The absolute and uniform convergence of this series is established in the same way as that in Theorem 6.13. Therefore repeated term by term differentiation of (6.2.4) is justified.

6.3 Paley–Wiener classes

Definition 6.15 *For $\sigma > 0$ and $1 \le p \le \infty$ we denote by PW_σ^p the Paley–Wiener class of functions f with a representation*

$$f(z) = \int_{-\sigma}^{\sigma} g(u)e^{izu}\,du, \qquad z \in \mathbb{C},$$

for some $g \in L^p(-\sigma,\sigma)$.

When $p = 2$ this is the definition given by Hardy (1941, p. 332) of a *Paley–Wiener function*. Functions of the class PW_σ^p belong to E_σ, because they are clearly holomorphic over the whole complex plane, and we have

$$|f(z)| \le \int_{-\sigma}^{\sigma} |g(u)|e^{-u\Im m z}\,du \le e^{\sigma|z|}\|g\|_1.$$

The inclusion relation $PW_\sigma^r \subset PW_\sigma^s$, $r > s \ge 1$, is an easy consequence of Hölder's inequality.

Again, it follows from Theorem 2.16 (whose conclusion remains true if $\mathcal{F}$ is replaced with $\mathcal{F}^{-1}$, as required here) that when $1 < p \le 2$, $f \in PW_\sigma^p$ implies that $f \in B_\sigma^q$, so that $PW_\sigma^p \subset B_\sigma^q$.

The opposite inclusion also holds when $p = q = 2$; this follows if we look forward for a moment to Theorem 7.2 of Paley–Wiener. Thus, we have $PW_\sigma^2 \equiv B_\sigma^2$. However, in spite of these relationships between Bernstein and Paley–Wiener classes, it will be convenient to treat each separately.

6.4   The cardinal series for Paley–Wiener classes

It is noticable that the hypotheses of Theorem 6.13 do not include the case $p \to \infty$ (this case will be investigated in Theorem 6.21). In the Paley–Wiener case it is at the other end of the scale of classes, that is, when $p = 1$, that something out of the ordinary occurs.

Theorem 6.16 *Let $f \in PW_{\pi w}^p$, $1 \le p \le \infty$. Then f is represented by its cardinal series uniformly on compact subsets of $\mathbb{C}$. When $p > 1$ the convergence is absolute. When $p = 1$ the convergence can fail to be absolute, but is absolute if g of Definition 6.15 belongs to $\mathfrak{Re}H^1$ (the definition of $\mathfrak{Re}H^1$ will be found in the next section, along with the proof of this part of the theorem).*

Proof   By definition,

$$f\left(\frac{n}{w}\right) = \frac{1}{\sqrt{2\pi w}} \int_{-\pi w}^{\pi w} g(u)e^{iun/w}\, du, \qquad n \in \mathbb{Z},$$

are the Fourier coefficients of g, so that

$$g(x) \sim \frac{1}{2\pi w} \sum_{n \in \mathbb{Z}} f\left(\frac{n}{w}\right) e^{-inx/w}.$$

Since $g \in L^p(-\pi w, \pi w) \cap L^1(-\pi w, \pi w)$, this formal Fourier series can be mutiplied by e^{itx} and integrated over $(-\pi w, \pi w)$. On the left we obtain

$$\int_{-\pi w}^{\pi w} g(x)e^{ixt}\, dx = f(t)$$

and on the right,

$$\frac{1}{2\pi w} \sum_{n \in \mathbb{Z}} f\left(\frac{n}{w}\right) \int_{-\pi w}^{\pi w} e^{ix(t-n/w)}\, dx = \sum_{n \in \mathbb{Z}} f\left(\frac{n}{w}\right) \operatorname{sinc}(wt - n),$$

as required.

Now let $1 < p \le 2$. From Hölder's inequality,

$$\sum_{n \in \mathbb{Z}} \left| f\left(\frac{n}{w}\right) \operatorname{sinc}(wt - n) \right| \le \left\{ \sum_{n \in \mathbb{Z}} \left| f\left(\frac{n}{w}\right) \right|^q \right\}^{1/q} \left\{ \sum_{n \in \mathbb{Z}} |\operatorname{sinc}(wt - n)|^p \right\}^{1/p}.$$

On the right the first of these sums converges by the Hausdorff–Young inequality for Fourier coefficients, Theorem 2.18, and the second converges uniformly on compact subsets of $\mathbb{C}$, as in the proof of Theorem 6.13. This completes the proof for the case $1 < p \leq 2$. For $p > 2$ we have $PW_{\pi w}^{p} \subset PW_{\pi w}^{2}$, so we can invoke the previous case.

When $p = 1$ the Hausdorff–Young inequality no longer applies, and to deal with this case we shall have to range a little further afield.

6.5   The space $\mathfrak{Re}H^{1}$

In this section we shall make a small excursion into the theory of H^{1} spaces. We need to do this in order to deal with the case $p = 1$ of Theorem 6.16; without this case the theorem would be seriously incomplete.

The general theory of H^{p} spaces is important in Fourier analysis. It is central to a general technique known as the use of *complex methods*, in which Fourier series representing periodic functions f are replaced with associated power series; such a power series represents an analytic function F associated with f. Powerful methods from the theory of analytic functions are used to study F, and the results are then re-interpreted as results about f. An excellent account of the rôle played by the space $\mathfrak{Re}H^{1}$ in these methods, including the proofs omitted here, is by Coifman and Weiss (1977, p. 569 ff.).

Definition 6.17 *The Hardy space H^{1} consists of functions F analytic in the unit disk $D = \{z \in \mathbb{C} : |z| < 1\}$ and for which*

$$\sup_{0 \leq r < 1} \int_{-\pi}^{\pi} |F(re^{i\theta})|\, d\theta < \infty.$$

Definition 6.18 $\mathfrak{Re}H^{1}$ *consists of the boundary values of the real parts of members of H^{1}; that is, if $F \in H^{1}$ then*

$$f(\theta) = \lim_{\substack{r \to 1 \\ r < 1}} \mathfrak{Re}F(re^{i\theta})$$

belongs to $\mathfrak{Re}H^{1}$.

We have the inclusion relations

$$L^{p}(-\pi, \pi) \subset \mathfrak{Re}H^{1} \subset L^{1}(-\pi, \pi), \qquad 1 < p \leq \infty.$$

It has often been noticed that there are results in Fourier analysis which hold for $L^{p}(-\pi, \pi)$, $p > 1$, fail to hold for all of $L^{1}(-\pi, \pi)$ and yet do hold for $\mathfrak{Re}H^{1}$. This, together with the inclusion relation above, prompted Coifman and Weiss to remark that "$\mathfrak{Re}H^{1}$ can be regarded as a good 'substitute' for $L^{1}(-\pi, \pi)$ that is endowed with many of the properties enjoyed by the L^{p} spaces for $1 < p < \infty$ and, moreover, contains all these spaces".

It is time to look at an example that illustrates this behaviour; it will also help in completing the proof of Theorem 6.16. Paley had proved that if $f \in L^{p}(-\pi, \pi)$, $1 < p \leq 2$, has the Fourier series $\sum_{n \in \mathbb{Z}} c_{n} e^{in\theta}$, then

$$\sideset{}{'}\sum_{n\in\mathbb{Z}} |c_n|^p |n|^{p-2} < \infty.$$

This result is not true in general for $f \in L^1(-\pi, \pi)$; this can be seen by considering the special series $\sum_{n=2}^{\infty} \cos n\theta / \log n$, which is known to be the Fourier series of an integrable function. Since $\sum_{n=2}^{\infty} 1/(n \log n)$ diverges, Paley's result (above) fails.

However, Hardy then showed that the extension of Paley's result to the case $p = 1$, that is

$$\sideset{}{'}\sum_{n\in\mathbb{Z}} \left| \frac{c_n}{n} \right| < \infty,$$

does hold if $f \in \mathfrak{Re} H^1$. This is called *Hardy's inequality* for $\mathfrak{Re} H^1$.

We now complete the proof of Theorem 6.16 (in the case $\sigma = \pi$; one passes to the general case by re-scaling). Let $\{c_n\}$ be the Fourier coefficients of $g \in L^1(-\pi, \pi)$ associated with $f \in PW_\pi^1$ by Definition 6.15. Then we have $f(n)/\sqrt{2\pi} = c_n$, and $f(n) = a_n$ in the inequality (1.1.10) of the Absolute Convergence Principle, Theorem 1.3. But this inequality can fail for the same example (above) which showed that Paley's result can fail for $f \in L^1$. On the other hand, Hardy's inequality and that in (1.1.10) are one and the same, so absolute convergence of the cardinal series holds when $g \in H^1$ by Theorem 1.3.

6.6 A convergence principle for Paley–Wiener spaces

In order to define the Hilbert function spaces known by the name Paley–Wiener (Definition 6.19 below), we take a slightly different approach to those taken hitherto. For example, we could have put $p = 2$ in Definition 6.15, but Definition 6.19 is more progressive. These two approaches are reconciled at the beginning of the next section.

We shall prove a convergence principle of some generality in the case of multi-dimensional band regions, although the multi-dimensional case will not be studied systematically until Chapter 14.

Let $m(B)$ denote N-dimensional Lebesgue measure of the set $B \subset \mathbb{R}^N$, and let B be the union of M disjoint components B_j, $j = 1, \ldots, M$, such that for every j, $0 < m(B_j) < \infty$.

Definition 6.19 *The Paley–Wiener space PW_B is defined by*

$$PW_B := \{f : f \in L^2(\mathbb{R}^N) \cap C(\mathbb{R}^N), \operatorname{supp} f^\wedge \subseteq \overline{B}\}. \tag{6.6.1}$$

If, in the one-dimensional case, B has two or more disjoint components, a member of PW_B is called a multi-band function.[1] If B is the single interval $[-\pi w, \pi w] \subset \mathbb{R}$, then we write $PW_{\pi w}$ for PW_B.

Using multi-dimensional versions of principles of Fourier analysis reviewed in §2.2 we shall prove that PW_B is a reproducing kernel Hilbert space, and that the

---

[1]This terminology was introduced by Landau (1967a, p. 1702).

reproducing kernel is uniformly bounded over $\mathbb{R}^N \times \mathbb{R}^N$. The main application that we shall make of this reproducing kernel theory is to the convergence of series in Paley–Wiener spaces, and a convergence principle will follow.

First, PW_B is a Hilbert space, whose inner product is just that of $L^2(\mathbb{R}^N)$. This follows from the unitary nature of the Fourier transform, expressed by Plancherel's theorem (Theorem 2.13 in multi-dimensional form). Thus, PW_B is isometrically isomorphic to $L^2(B)$, a relationship that we shall often refer to as *Fourier duality*. In particular, this means that every basis for $L^2(B)$ has its counterpart in PW_B via inverse Fourier transformation. Indeed, the property of being a basis, an orthonormal basis or a Riesz basis is possessed by $\{\varphi_n\}$ in $L^2(B)$ if and only if it is possessed by $\{\mathcal{F}^{-1}\varphi_n\}$ in PW_B.

When $f \in PW_B$ we have

$$f(t) = \frac{1}{(2\pi)^{N/2}} \int_B f^\wedge(x) e^{ix \cdot t} \, dx \qquad \text{a.e.} \qquad (6.6.2)$$

where t and x are N-dimensional multi-variables, and where $x \cdot t$ denotes the scalar product. This follows from Theorem 2.15 in multi-dimensional form; by continuity the representation (6.6.2) actually holds for *all* $t \in \mathbb{R}^N$. Hence, by the "tautology trick" of making the implied characteristic function explicit, we have

$$f(t) = \frac{1}{(2\pi)^{N/2}} \int_{\mathbb{R}^N} f^\wedge(x) \chi_B(x) e^{ix \cdot t} \, dx, \qquad t \in \mathbb{R}^N, \qquad (6.6.3)$$

so that by Schwarz' inequality applied in $L^2(\mathbb{R}^N)$,

$$|f(t)| \le \frac{m(B)}{(2\pi)^{N/2}} \|f^\wedge\| = (\text{constant}) \|f\|.$$

This means that the "point evaluation" functional is bounded (and obviously linear) on PW_B, and it follows from property 1 of §3.4 that PW_B is an rk space.

Now, by Plancherel's theorem in conjunction with (6.6.3) we have

$$f(t) = \frac{1}{(2\pi)^{N/2}} \int_{\mathbb{R}^N} f(s) \overline{\mathcal{F}^{-1} \left(\chi_B(\cdot) e^{-i(\cdot) \cdot t} \right)(s)} \, ds. \qquad (6.6.4)$$

This is the reproducing formula for PW_B, where the reproducing kernel is given by

$$k(s,t) = \overline{\mathcal{F}^{-1} \left(\chi_B(\cdot) e^{-i(\cdot) \cdot t} \right)(s)} = \frac{1}{(2\pi)^N} \int_B e^{ix \cdot (s-t)} \, dx.$$

Hence, by a simple estimate,

$$|k(s,t)| \le \frac{m(B)}{(2\pi)^N}, \qquad s, t \in \mathbb{R}^N,$$

and since this estimate holds in particular when $s = t$, the criterion in property 4 of §3.4 holds. We therefore have the following:

CONVERGENCE PRINCIPLE. Let $B \subset \mathbb{R}^N$ be a multi-band region, and let $f \in PW_B$. If an expansion for f converges in the norm of PW_B, then it also conveges pointwise and globally uniformly over $\mathbb{R}^N$ (that is, uniformly over all of $\mathbb{R}^N$, not merely on compact subsets).

Looking forward for a moment to the Paley–Wiener theorem of §7.1, we find that, in one dimension, we can extend the convergence in this principle to hold pointwise on compact subsets of the complex plane $\mathbb{C}$. This is because, given any compact subset, analytic continuation of the result is possible into a region of $\mathbb{C}$ containing that compact subset.

6.7 Ordinary Paley–Wiener space and its reproducing kernel

The Paley–Wiener classes PW_σ^p were defined in §6.3. In this section we shall take $p = 2$ and $\sigma = \pi w$, so that PW_σ^2 consists of those functions f for which

$$f(t) = \int_{-\pi w}^{\pi w} g(x)e^{ixt}\, dx, \qquad g \in L^2(-\pi w, \pi w). \qquad (6.7.1)$$

It is usual to omit the 2 from the notation and refer simply to the ordinary Paley–Wiener space $PW_{\pi w}$. Several special examples of Paley–Wiener functions are to be found in Appendix A, nos. 8-22.

Equation (6.7.1) and Definition 6.19 give ostensibly different definitions of $PW_{\pi w}$; in fact the definitions are equivalent. To see this we have, following Definition 6.19 and using Theorem 2.15,

$$f(t) = \lim_{A \to \infty} \frac{1}{\sqrt{2\pi}} \int_{-\pi w}^{\pi w} \left(1 - \frac{|x|}{A}\right) f^\wedge(x)e^{ixt}\, dx$$

$$= \frac{1}{\sqrt{2\pi}} \int_{-\pi w}^{\pi w} f^\wedge(x)e^{ixt}\, dx, \qquad \text{a.e.,} \qquad (6.7.2)$$

and f is almost everywhere equal to a continuous function. Now the requirement $f \in C(\mathbb{R})$ picks out the continuous member of each equivalence class, the representation above holds now for all $t \in \mathbb{R}$, and in (6.7.1) g can be replaced with $f^\wedge/\sqrt{2\pi}$. In this way our two seemingly different definitions of $PW_{\pi w}$ are reconciled.

We now need to look at the actual form of the reproducing kernel for $PW_{\pi w}$. Suppose first that $\{h_n(x)\}$, $n \in \mathbb{X}$, is an orthonormal basis for $L^2(-\pi w, \pi w)$, and that we expand the Fourier kernel e^{-ixt} in this set. Then from Parseval's relation we have

$$\langle e^{-i\cdot t}, e^{-i\cdot s}\rangle = \sum_{n \in \mathbb{X}} \langle e^{-i\cdot t}, h_n\rangle \overline{\langle e^{-i\cdot s}, h_n\rangle}, \qquad (6.7.3)$$

in which each inner product is that of $L^2(-\pi w, \pi w)$. The left-hand side is just

$$\int_{-\pi w}^{\pi w} e^{i(s-t)v}\, dv = 2\pi w \operatorname{sinc} w(s - t),$$

and the inner products on the right-hand side represent $\sqrt{2\pi}\,\overline{h_n^\vee(t)}$ and $\sqrt{2\pi}\,h_n^\vee(s)$ respectively. Hence (6.7.3) can be written

$$w\operatorname{sinc}w(s-t) = \sum_{n\in X} h_n^\vee(s)\,\overline{h_n^\vee(t)}. \tag{6.7.4}$$

But since $\{h_n^\vee\}$ is, by Plancherel's theorem, an orthonormal basis for $PW_{\pi w}$, the right-hand side of (6.7.4) is a series of the kind in property 3 of §3.4, and therefore the left-hand side shows that $w\operatorname{sinc}w(s-t)$ is the reproducing kernel for $PW_{\pi w}$. Therefore *the reproducing equation for $f \in PW_{\pi w}$*, because of condition (b) of §3.4, is

$$f(t) = w\int_{\mathbb{R}} f(s)\operatorname{sinc}w(s-t)\,ds. \tag{6.7.5}$$

By a simple change of variable this equation can be re-written as a precise integral analogue of the cardinal series as in Lemma 6.20 below.

Of course, the series (6.7.4) takes on a different form for each choice of the orthonormal basis $\{h_n\}$, and we cannot leave this topic without looking at a few of the possibilities.

An obvious choice is $h_n(x) = e^{-ixn/w}/\sqrt{2\pi w}$. This gives

$$\operatorname{sinc}w(s-t) = \sum_{n\in\mathbb{Z}} \operatorname{sinc}(ws-n)\operatorname{sinc}(wt-n). \tag{6.7.6}$$

This is simply the cardinal series for the kernel as a function of s for each real t (or of t for each real s).

A more interesting choice is to take $h_n(x) = (n+\frac{1}{2})^{1/2}P_n(x)$, $n \in \mathbb{N}_0$, where P_n is the nth Legendre polynomial (defined in Example 4.2), which is known to be an orthonormal basis for $L^2(-1,1)$ (here we have taken $\pi w = 1$). Using the Fourier transform (Appendix A, no. 19)

$$\left(\mathcal{F}i^n t^{-\frac{1}{2}}J_{n+\frac{1}{2}}(t)\right)(x) = P_n(x)\chi_{[-1,1]}(x), \qquad n \in \mathbb{N}_0,$$

where $J_{n+\frac{1}{2}}$ is the Bessel function of half odd integer order, we obtain

$$\frac{\sin(s-t)}{(s-t)} = \pi\sum_{n=0}^{\infty}(n+\tfrac{1}{2})\frac{J_{n+\frac{1}{2}}(s)}{s^{\frac{1}{2}}}\frac{J_{n+\frac{1}{2}}(t)}{t^{\frac{1}{2}}}.$$

This beautiful formula dates from the great age of formulae in the nineteenth century (e.g. Nielsen 1904, pp. 278 ff.).

Also it follows that $\{[(n+\frac{1}{2})/t]^{1/2}J_{n+\frac{1}{2}}(t)\}$, $n \in \mathbb{N}_0$, is an orthonormal basis for PW_1. This expansion set is associated with the Bessel–Neumann series (Higgins 1972, p. 712).

We are now going to collect together several facts that have emerged from our foregoing discussion; most of them are due to Hardy (1941). Hardy did not use phrases such as "Hilbert space", or "reproducing kernel", but the essential features of the following lemma are present in his classic treatment of the cardinal series.[2]

Lemma 6.20

1. The Paley–Wiener space $PW_{\pi w}$ is a Hilbert space with reproducing kernel. The reproducing kernel is $w\operatorname{sinc}w(s-t)$, $(s,t)\in\mathbb{R}\times\mathbb{R}$, and the reproducing equation for $f\in PW_{\pi w}$ can be written

$$f(t) = \int_{\mathbb{R}} f(\frac{v}{w})\operatorname{sinc}(wt-v)\,dv.$$

2. If f belongs to $PW_{\pi w}$ we have the representation

$$f(t) = \frac{1}{\sqrt{2\pi}}\int_{-\pi w}^{\pi w} f^\wedge(x)e^{ixt}\,dx.$$

Such a function f is bounded on $\mathbb{R}$, and $f(t) \to 0$ as $t \to \pm\infty$. If $f^\wedge \in BV(-\pi w, \pi w)$ then $f = \mathcal{O}(1/t)$, $|t| \to \infty$.
3. The set $\{\sqrt{w}\operatorname{sinc}(wt-n)\}$, $n \in \mathbb{Z}$, is an orthonormal basis for $PW_{\pi w}$, with respect to which the n th Fourier coefficient of f is $(1/\sqrt{w})f(n/w)$. The cardinal series for $f \in PW_{\pi w}$ is the orthogonal expansion

$$f(t) = \sum_{n\in\mathbb{Z}} f(\frac{n}{w})\operatorname{sinc}(wt-n),$$

which converges in the norm of $PW_{\pi w}$, and absolutely and globally uniformly over $\mathbb{R}$. The expansion also holds uniformly on compact subsets of the complex t-plane.
4. Parseval's relation for $f \in PW_{\pi w}$ is

$$\|f\|^2 = \frac{1}{w}\sum_{n\in\mathbb{Z}}\left|f\left(\frac{n}{w}\right)\right|^2.$$

To complete the proof we note first that the second part of item 2 follows from Theorem 2.6 of Riemann–Lebesgue, since $f^\wedge$ belongs to $L^2(-\pi w, \pi w)$ and hence to $L^1(-\pi w, \pi w)$.

We need only to refer to Theorem 6.16 for the facts concerning absolute convergence, and to the Convergence Principle of the previous section, and the remark following it, for the uniform convergence.

---

[2]The first explicit introduction of Hilbert space methods into the theory appears to be by Weston (1949) (a commentary and further references can be found in Higgins 1985, p. 58), and independently by Beutler (1966). The presence of reproducing kernels was first pointed out by Yao (1967), and independently by Higgins (1972). The connection between reproducing kernels, bases and their duals in this context was emphasized by Higgins (1972, 1977).

As to part 4, it follows from the analysis of this section that, for $n \in \mathbb{Z}$,

$$\left(\frac{e^{-ixn/w}}{\sqrt{2\pi w}} \chi_{[-\pi w, \pi w]} \right)^{\vee} (t) = \sqrt{w} \operatorname{sinc} w \left(t - \frac{n}{w} \right)$$

are the members of an orthonormal basis for $PW_{\pi w}$. The Fourier coefficients for f with respect to this set are

$$\sqrt{w} \int_{\mathbb{R}} f(s) \operatorname{sinc} w \left(s - \frac{n}{w} \right) ds = \frac{1}{\sqrt{w}} f \left(\frac{n}{w} \right),$$

using the reproducing equation. It remains to appeal to Theorem 3.2, and the proof is now complete.

6.8 Sampling and entire functions of polynomial growth

By the inclusions in Lemma 6.6, $f \in B_{\sigma}^p$, $p \geq 1$, is bounded over $\mathbb{R}$. Hence, by Theorem 7.7 the Fourier transform of f is a distribution with support in $[-\sigma, \sigma]$. We can say that Bernstein functions are *band-limited in the distributional sense.* It makes sense, therefore, to retain the notion of the Nyquist sampling rate (Definition 1.4) for Bernstein spaces, and naturally we take it to be σ/π.

We have seen that if f belongs to B_{σ}^p, $1 \leq p < \infty$, then it can be represented by its cardinal series with sampling at the Nyquist rate. If we pass to the case $f \in B_{\sigma}^{\infty}$ this is no longer true in general; $\sin \pi t$ belongs to B_{π}^{∞} and is zero at every sample point, so the expansion breaks down. It does not do so for every $f \in B_{\sigma}^{\infty}$ however, as we shall see in Example 6.23.

Nevertheless, the question remains as to whether there is an extension of the sampling ideas which worked for B_{σ}^p when p was finite that will always work when $p \to \infty$. There are at least two ways of doing this; they both involve extracting more information from f than was required hitherto. In one of them we can retain the Nyquist sampling rate and then adjoin more information; it seems quite remarkabe that just one more sample, $f'(0)$, is sufficient. The other is to increase the overall sampling rate; an arbitrarily small increase is sufficient. These ideas are embodied in the following theorem; the series in part (a) is often called Tschakaloff's series, and that in (b) Zakai's series. Comments on these series, with further references, can be found in Higgins (1985, p. 54) and Zayed (1993, p. 31).

Theorem 6.21 *Let $f \in B_{\pi}^{\infty}$. Then for each real t we have*

$$(a) \qquad f(t) = \frac{\sin \pi t}{\pi} \left\{ f'(0) + \frac{f(0)}{t} + t \sum_{n \in \mathbb{Z}}' f(n) \frac{(-1)^n}{n(t-n)} \right\}; \qquad (6.8.1)$$

$$(b) \qquad f(t) = \sum_{n \in \mathbb{Z}} f(n/w) \operatorname{sinc}(wt - n), \qquad w > 1. \qquad (6.8.2)$$

The series in (a) converges absolutely for each real t; that in (b) may fail to do so.

Proof We shall need the well known partial fractions expansion

$$\pi \csc \pi wt - \frac{1}{wt} = \sum_{n\in\mathbb{Z}}{}'(-1)^n \left(\frac{1}{wt-n} + \frac{1}{n} \right). \tag{6.8.3}$$

Let us define an auxilliary function g by taking

$$\begin{cases} g(t) = \dfrac{f(t) - f(0)}{t}, & t \neq 0; \\[2mm] g(0) = f'(0). \end{cases}$$

Clearly g belongs to $L^2(\mathbb{R})$ and has the same exponential type as f, therefore by the Paley–Wiener theorem (Theorem 7.2), $g \in PW_{\pi w}$, $w \geq 1$.

Next we note that (6.8.1) and (6.8.2) are both true when $t = 0$. Now g can be expanded in cardinal series, so that we have

$$g(t) = \frac{f(t) - f(0)}{t}$$

$$= f'(0)\,\text{sinc}\,wt + \sum_{n\in\mathbb{Z}}{}' \frac{f(n/w) - f(0)}{n/w}\,\text{sinc}(wt - n). \tag{6.8.4}$$

To pursue our first alternative we must take $w = 1$. Now that part of the sum in (6.8.4) containing $f(0)$ can be written in the form

$$-f(0)\frac{\sin \pi t}{\pi t}\sum_{n\in\mathbb{Z}}{}'(-1)^n \left(\frac{1}{t-n} + \frac{1}{n} \right),$$

so that, using (6.8.3) with $w = 1$, we can write (6.8.4) in the form

$$f(t) = f(0) + f'(0)\frac{\sin \pi t}{\pi}$$

$$+ t\sum_{n\in\mathbb{Z}}{}' f(n)\frac{\sin \pi(t-n)}{n\pi(t-n)} - f(0)\frac{\sin \pi t}{\pi}\left[\pi \csc \pi t - \frac{1}{t} \right], \tag{6.8.5}$$

and a little further reduction gives (6.8.1).

The absolute convergence of (6.8.1) for each real t is assured, since f is bounded over $\mathbb{R}$.

To prove part (b) we need a lemma which is of some interest in its own right; for example, it is instructive to compare it with the series in Problem 6.5.

Lemma 6.22 Let $g \in PW_\pi$, so that

$$g(t) = \int_{-\pi}^{\pi} g^\wedge(x)e^{ixt}\,dx.$$

Then

$$\sum_{n\in\mathbb{Z}}(-1)^n g(n/w) = 0, \qquad w > 1.$$

Proof First, we have $g^\wedge \in L^2(-\pi,\pi) \cap L^1(-\pi,\pi)$. Now, in the standard summation formula

$$\sum_{n=-N_1}^{N_2} (-1)^n e^{inx/w} = \frac{(-1)^{N_1}e^{-ixN_1/w} + (-1)^{N_2}e^{ix(N_2+1)/w}}{1 + e^{ix/w}},$$

the condition $w > 1$ ensures that $1 + e^{ix/w}$ is never zero when $x \in [-\pi,\pi]$. Hence

$$\sum_{n=-N_1}^{N_2}(-1)^n g(n/w) = \sum_{n=-N_1}^{N_2}(-1)^n \int_{-\pi}^{\pi} g^\wedge(x)e^{ixn/w}\,dx$$

$$= \int_{-\pi}^{\pi}\left(\frac{g^\wedge(x)}{1+e^{ix/w}}\right)\left[(-1)^{N_1}e^{-ixN_1/w} + (-1)^{N_2}e^{ix(N_2+1)/w}\right]dx.$$

Now, because $g^\wedge(x)/(1 + e^{ix/w}) \in L^1(-\pi,\pi)$, the Riemann–Lebesgue Theorem applies as $N_1, N_2 \to \infty$ and this completes the proof.

Now we can complete the proof of part (b) of Theorem 6.21. First the sum in (6.8.4) can be written in the form

$$\frac{1}{t}\sum_{n\in\mathbb{Z}}{}'[f(n/w) - f(0)]\,\text{sinc}(wt-n) + \text{sinc}\,wt\sum_{n\in\mathbb{Z}}{}'(-1)^n\frac{f(n/w) - f(0)}{n/w}.$$

In the first of these two sums we can replace $\sum'\text{sinc}(wt-n)$ with $1 - \text{sinc}\,wt$, since this just involves a rearrangement of (6.8.3), after noticing that the second summand on the right-hand side gives rise to a vanishing sum. The second sum reduces to $-f'(0)$ by Lemma 6.22. Now (6.8.4) becomes

$$f(t) = f(0) + t\Bigg\{f'(0)\,\text{sinc}\,wt$$

$$+ \frac{1}{t}\left[\sum_{n\in\mathbb{Z}}{}'f(n/w)\,\text{sinc}(wt-n) - f(0)(1 - \text{sinc}\,wt)\right] - f'(0)\,\text{sinc}\,wt\Bigg\}$$

and a little further rearrangement gives (6.8.2).

That (6.8.2) may fail to converge absolutely is illustrated in the following:

Example 6.23 Let $f(t) = e^{iat}$, $-\pi < a < \pi$; then $f \in B_\pi^\infty$. The series (6.8.2) gives

$$e^{iat} = \sum_{n\in\mathbb{Z}}e^{ian/w}\,\text{sinc}(wt-n)$$

for every $t \in \mathbb{R}$, and $w > 1$. When wt is not an integer the series is clearly not absolutely convergent.

It is remarkable that this series representation still holds even when $w = 1$. This is because it is then the Fourier series of e^{iat} regarded as a function of a, as the reader will easily be able to check. So the Nyquist sampling rate can in fact be achieved for this example.

Again, we have

$$e^{iat} = \int_{-\pi}^{\pi} \delta(x - a)e^{ixt}\, dx,$$

which shows that $\delta(x - n)$ is the Fourier transform of e^{iat} in the sense of distributions.

This example was given by Hardy and by Campbell from two rather different points of view. Further references can be traced in Higgins (1985, p. 71).

$$* \qquad * \qquad *$$

It is desirable to go beyond the Bernstein spaces and look for a sampling theory that will apply to functions which are not bounded on $\mathbb{R}$. There are two main approaches to this problem. One is to extend the procedure by which we proved Theorem 6.21, part (a), by subtracting more terms of the Talor series for f. The next stage in this process appears in Problem 6.9, but here it is difficult to proceed in general.

A different approach is via the theory of distributions. Some background on distributions will be found in §2.4. The previous example saw a function that was band-limited in the sense of distributions; another example is found in the following:

Example 6.24 We consider the *sine integral function* (e.g. Magnus *et al.* 1966, p. 347), which can be written in the form

$$\mathrm{Si}(t) = \frac{1}{2i}\mathrm{P.V.} \int_{-1}^{1} \frac{e^{iut}}{u}\, du$$

(P.V. indicates a principle value integral). The function $\mathrm{Si}(t)$ occurs naturally in applications in that it represents the response of an ideal low-pass filter to a step function. It can be thought of as the Fourier transform of a "principle value distribution"; a distribution with compact support (see Definition (7.5)).

Example 6.25 Let δ be replaced by δ' in Example 6.23. The reader can easily show that the cardinal series for ite^{iat} diverges.

With these examples in mind we adopt the following:

Definition 6.26 *Let $r \in \mathbb{N}$. The class of functions f such that*

$$\int_{\mathbb{R}} \frac{|f(t)|}{(1+t^2)^r}\, dt < \infty,$$

and such that f is band-limited to $[-\pi w, \pi w]$ in the sense of distributions, that is, $f^\wedge$ is a tempered distribution with support in $[-\pi w, \pi w]$, is denoted by $B_{r,\pi w}$.

The Theorem 7.7 of Paley–Wiener–Schwartz shows that members of $B_{r,\pi w}$ can be extended to the complex plane as entire functions of exponential type at most πw, and that they are polynomially bounded along the real axis.

Theorem 6.27 *Let $f \in B_{r,\pi w}$ for some $r \in \mathbb{N}$. Then*

$$f(t) = P_r(t) \cos \pi w t + \sum_{n \in \mathbb{Z}} \left(\frac{wt}{n}\right)^r \left\{ f\left(\frac{n}{w}\right) - P_r\left(\frac{n}{w}\right)(-1)^n \right\} \operatorname{sinc}(wt - n),$$

where

$$P_r(t) := \sum_{s=0}^{r-1} \frac{t^s}{s!} \left[\left(\frac{d}{d\tau}\right)^s \frac{f(\tau)}{\cos \pi w \tau} \right]_{\tau=0}$$

consists of the first r terms of the formal Taylor series for $f(t)/\cos \pi w t$, in which the term corresponding to $n = 0$ is taken to be

$$\frac{t^r}{r!} \left[\left(\frac{d}{d\tau}\right)^r \frac{f(\tau)}{\cos \pi w \tau} \right]_{\tau=0} \operatorname{sinc} wt.$$

The expansion holds uniformly on compact subsets of $\mathbb{R}$.

The proof of this theorem is beyond our present scope; it can be found, in a slightly more general form, in Hoskins and Sousa Pinto (1994, p. 130).

PROBLEMS

6.1 Show that $\operatorname{sinc} t$ does not belong to $L^1(\mathbb{R})$. (Hint: write down its L^1-norm, and then estimate the integrand by triangles.)

Comment on the fact that the cardinal series expansion holds for $f \in B_\sigma^1$ (by Theorem 6.13), but that the expansion functions do not belong to B_σ^1.

Obtain the improper Riemann integral

$$\int_0^\infty \frac{\sin at}{t}\, dt = \frac{\pi}{2} \operatorname{sgn} a.$$

6.2 Show that
$$\frac{\operatorname{sinc} z}{\operatorname{sinc} \sqrt{z} \operatorname{sinc} \sqrt{-z}}, \qquad z \in \mathbb{C}$$
is a Paley–Wiener function. This special function features in a proof by Seip (1995, p. 155).

6.3 Show that if $f \in PW_{\pi w}$ and is real valued on $\mathbb{R}$, then f is orthogonal to its Hilbert transform $f^{\sim}$ and to its derivative f'; and in fact to all its derivatives of odd order.

6.4 Show that a more general and more symmetric form of Poisson's summation formula is
$$\sqrt{A} e^{\pi i a b} \sum_{n \in \mathbb{Z}} e^{2\pi i n a} f(A(n+b)) \sim \sqrt{B} e^{-\pi i a b} \sum_{n \in \mathbb{Z}} e^{2\pi i n b} f^{\wedge}(B(n-a)).$$

Here, A, B, a and b are real, and $AB = 2\pi$. Boas (1972, p. 122) recalls that Hardy used to give this form in lectures.

6.5 Let $f \in B_\pi^1$. By taking $a = \frac{1}{2}$, $b = 0$ and $A = 1$ in Problem 6.4, obtain the summation formula
$$\sum_{n \in \mathbb{R}} (-1)^n f(n) = 0.$$

6.6 Show that, for $f \in B_\pi^1$,
$$\sqrt{2} \sum_{n \in \mathbb{R}} (-1)^n f(2n) = \sqrt{\pi} [f^{\wedge}(\pi/2) + f^{\wedge}(-\pi/2)],$$
$$\sqrt{3} \sum_{n \in \mathbb{R}} f(3(n-a)) = \sqrt{2\pi/3} [f^{\wedge}(0) + e^{2\pi i a} f^{\wedge}(2\pi/3) + e^{-2\pi i a} f^{\wedge}(-2\pi/3)]$$

(Boas (1972) contains more remarkable formulae of this kind.)

6.7 Show that $\cos \pi t$ belongs to B_π^∞ and has a cardinal series expansion.

6.8 Show that for every $f \in B_{\pi w}^p$, $1 \le p < \infty$, the representations
$$f'(t) = \pi w \sum_{n \in \mathbb{Z}} f\left(\frac{n}{w}\right) \left\{ \frac{\cos \pi(wt-n)}{\pi(wt-n)} - \frac{\sin \pi(wt-n)}{(\pi(wt-n))^2} \right\},$$
$$f'(t) = w \sum_{n \in \mathbb{Z}}' f\left(t + \frac{n}{w}\right) \frac{(-1)^{n+1}}{n},$$
$$f'(t) = \frac{4w^2}{\pi} \sum_{n \in \mathbb{Z}} f\left(t + \frac{2n+1}{2w}\right) \frac{(-1)^n}{(2n+1)^2}$$

hold, with uniform convergence on compact subsets of $\mathbb{R}$ (Butzer *et al.* 1988, p. 13).

6.9 Show that if $f \in E_\pi$ and is $\mathcal{O}(|x|)$, $|x| \to \infty$, then

$$f(t) = \frac{\sin \pi t}{\pi} \left\{ \frac{t}{2} f''(0) + f'(0) + \left(\frac{1}{t} + \frac{t\pi^2}{6} \right) f(0) + t^2 \sum_{n \in \mathbb{Z}}{}' f(n) \frac{(-1)^n}{n^2(t-n)} \right\}.$$

Follow the same procedure as in the proof of Theorem 6.21, part (a), but this time start by subtracting from f the first two terms of its Taylor series.

6.10 Let α and β be real, and $0 < \gamma \leq 1$. Show that

$$\sum_{n \in \mathbb{Z}} \operatorname{sinc} \gamma(\alpha - n) \operatorname{sinc}(\beta - n)$$

$$= \int_{\mathbb{R}} \operatorname{sinc} \gamma(\alpha - v) \operatorname{sinc}(\beta - v)\, dv = \operatorname{sinc} \gamma(\alpha - \beta).$$

6.11 Show that if $0 < \alpha < \pi$, and ν is either 1 or 2, then we have the series/integral analogues: [3]

$$\sum_{n \in \mathbb{Z}} \left(\frac{\sin(n\alpha + \theta)}{n\alpha + \theta} \right)^\nu = \int_{\mathbb{R}} \left(\frac{\sin(v\alpha + \theta)}{v\alpha + \theta} \right)^\nu dv = \frac{\pi}{\alpha}.$$

---

[3] A discussion of this interesting topic, with further references, can be found in Boas and Pollard (1973). They point out (p. 19) that "Continuous analogues of series are of interest in physics ... where one often attempts to deal with an intractable sum by replacing it by the corresponding integral. In fact, the sum (the case $\nu = 2$ above) ... was "approximated" by the integral before it was realized that the approximation is exact."

7

MORE ABOUT PALEY–WIENER SPACES

The four topics covered in this chapter are not directly concerned with sampling, but with various aspects of Paley–Wiener spaces. Since these spaces are typically collections of band-limited functions, and the notion of band-limitation in one form or another runs through all our work on sampling, this material has been included to provide some necessary background, and to lend perspective on these important spaces.

7.1 Paley–Wiener theorems — a review

For general reference it will be convenient to have at our disposal a collection of Paley–Wiener theorems. This kind of theorem serves the very important purpose of linking Fourier analytic methods on the one hand with the methods of complex function theory on the other. Again for general reference we have:

Definition 7.1 *The support of $F \in L^p(\mathbb{R}^N)$, $p \geq 1$, written* supp F, *is the set of all points $x \in \mathbb{R}^N$ that possess no neighbourhood throughout which $F(x) = 0$ a.e.*

If S denotes the support of F, then S is a closed set; hence it is compact if it is bounded. One then speaks of F as being "compactly supported on S".

The original theorem of Paley and Wiener (1934, p. 12) runs as follows:

The class of functions $\{f(z)\}$ whose members are entire, belong to L^2 when restricted to the real axis, and are such that $|f(z)| = o\left(e^{\sigma|z|}\right)$, is identical to the class of functions whose members have a representation

$$f(z) = \int_{-\sigma}^{\sigma} \varphi(u)e^{iuz}\,du, \qquad (7.1.1)$$

for some $\varphi \in L^2(-\sigma, \sigma)$.

Paley and Wiener used the condition (6.1.3), the most restrictive of the three conditions associated with the notion of exponential type. However, in the present context this restrictiveness is apparent, not actual. In fact, the theorem holds under the condition (6.1.1).

The integral representation (7.1.1) suggests that φ is, up to a constant multiple, the Fourier transform of f; because of the inversion of Theorem 2.15 this is indeed the case, and prompts us to restate the theorem in a form which is slightly more useful for our present needs (a proof will be found in §9.1).

Theorem 7.2 (Paley–Wiener Theorem) *Let $f \in L^2(\mathbb{R})$. Then f has an analytic extension to $\mathbb{C}$ which belongs to E_σ if and only if $\operatorname{supp} f^\wedge \subseteq [-\sigma, \sigma]$.*

It should be observed that the hypotheses do not require σ or $-\sigma$ to belong to $\operatorname{supp} f^\wedge$; if either does belong to $\operatorname{supp} f^\wedge$, then we can conclude that the exponential type of f is *equal* to σ.

Theorem 7.3 (Paley–Wiener Theorem — L^p version)
(a) If a function f belongs to E_σ and its restriction to $\mathbb{R}$ belongs to $L^p(\mathbb{R})$, $1 < p < 2$, then there exists $\varphi \in L^q(-\sigma, \sigma)$, $q = p/(p-1)$, such that

$$f(z) = \int_{-\sigma}^{\sigma} e^{izu} \varphi(u)\, du.$$

(b) If $\varphi \in L^p(-\sigma, \sigma)$, $1 < p < 2$, and

$$f(z) = \int_{-\sigma}^{\sigma} e^{izu} \varphi(u)\, du,$$

then f belongs to E_σ and its restriction to $\mathbb{R}$ belongs to $L^q(\mathbb{R})$.

The case $p = 1$ calls for a slightly different version of this theorem.

Theorem 7.4 (Paley–Wiener Theorem — L^1 version) *A function f belongs to E_σ, and when restricted to $\mathbb{R}$ belongs to $L^1(\mathbb{R})$, if and only if it has a representation*

$$f(z) = \int_{-\sigma}^{\sigma} e^{izu} \varphi(u)\, du$$

where $\varphi(\sigma) = \varphi(-\sigma) = 0$, and φ extended as being 0 outside $(-\sigma, \sigma)$ has an absolutely convergent Fourier series on $(-\sigma - \delta, \sigma + \delta)$, $\delta > 0$.

A comprehensive discussion of these L^p Paley–Wiener theorems can be found in Boas (1954, §6.8). It will also be convenient to include here a distributional form of the Paley–Wiener theorem. This theorem and its proof can be found in the book of Hoskins and Sousa Pinto (1994, p. 118). First, we need two definitions.

Definition 7.5 *The distribution φ is said to vanish on $A \subset \mathbb{R}^N$ if $\varphi(f) = 0$ for any test function f with $\operatorname{supp} f \subset A$. The support of a distribution φ is the complement of the union of all open sets on which φ vanishes. The support of φ is denoted by $\operatorname{supp}\varphi$; it is of course a closed set.*

Definition 7.6 *By P_σ we denote the class of functions that belong to E_σ, and are bounded by some polynomial on the real axis.*

Theorem 7.7 (Paley–Wiener–Schwartz Theorem) *If f belongs to P_σ then it has a Fourier transform in the sense of distributions, and this Fourier transform is a distribution with compact support in $[-\sigma, \sigma]$.*[1] *Conversely, if f is the*

---

[1] Such a distribution is tempered (Hoskins and Sousa Pinto 1994, p. 112), so the class of tempered distributions, outlined in §2.4, remains the appropriate one here.

inverse Fourier transform of a distribution with compact support in $[-\sigma, \sigma]$, *then* f *belongs to* P_σ.

There are several other types of Paley–Wiener theorem, the most relevant to our present purposes being the multi-dimensional forms. Here we refer the reader to the books of Stein and Weiss (1971) and Nikol'skiĭ (1975); also, multi-dimensional versions in an interesting applied setting can be found in Smith *et al.* (1977).

7.2 Bases for Paley–Wiener spaces

We saw in §6.6 that $\{\varphi_n^\vee\}$ is an orthonormal basis for $PW_{\pi w}$ if $\{\varphi_n\}$ is an orthonormal basis for $L^2(-\pi w, \pi w)$. Two of the possibilites occurred in §6.7. One of them was to take the ordinary exponential Fourier functions as orthonormal basis for $L^2(-\pi w, \pi w)$; then the "Fourier duality" argument gave rise to the translated "sinc" functions of the cardinal series as orthonormal basis of $PW_{\pi w}$ (this was the original argument of Hardy (1941)). Again, in just the same way the Legendre polynomials gave rise to a set of Bessel–Neumann functions as orthonormal basis for PW_1 (although here the series is not a sampling series). In this section a few more examples along these lines will be investigated.

Lemma 7.8 *Let* $\{\varphi_n\}$, $n \in X$, *be an orthonormal basis for* $L^2(-\pi w, \pi w)$. *Then so is* $\{\omega\, \varphi_n\}$, *where* ω *is any measurable function defined on* $[-\pi w, \pi w]$ *such that* $|\omega| \equiv 1$.

Proof For the completeness, $\langle \omega\, \varphi_n, f \rangle = 0$, $n \in X$, implies that $\overline{\omega}\, f$, and hence f, is null. Since $|\omega|^2 \equiv 1$ the orthonormality follows by integrating $\omega\varphi_n\, \overline{\omega\varphi_m} = \varphi_n\, \overline{\varphi_m}$ over $[-\pi w, \pi w]$.

Corollary 7.9 *In Lemma 7.8 take* $\omega(x) = e^{-i\psi(x)}$ *and* $\varphi_n(x) = e^{-inx/w}/\sqrt{2\pi w}$. *Let*

$$T(t) := \frac{1}{\sqrt{2\pi w}} \mathcal{F}^{-1}\left(\chi_{\pi w} e^{-i\psi}\right)(t).$$

Then $\{T(t - n/w)\}$, $n \in \mathbb{Z}$, *is an orthonormal basis for* $PW_{\pi w}$.

Proof By Lemma 7.8,

$$\left\{ \frac{1}{\sqrt{2\pi w}} e^{-i\psi(x) - inx/w} \right\}, \qquad n \in \mathbb{Z},$$

is an orthonormal basis for $L^2(-\pi w, \pi w)$. By Fourier duality the nth member of an orthonormal basis for $PW_{\pi w}$ is

$$\frac{1}{\sqrt{2\pi w}} \mathcal{F}^{-1}\left[\chi_{\pi w} e^{-i\psi(\cdot)} e^{-in\cdot/w}\right](t) = T(t - n/w).$$

using Appendix A, no. 3a.

A simple example is obtained by taking $\psi(x) = ax$. This leads to a "shifted" cardinal series, derived in a slightly different way in Example 8.3.

Example 7.10 In Corollary 7.9 take $w = 1$ for convenience, and

$$\psi(x) = \begin{cases} ax, & -\pi \le x < 0; \\ bx, & 0 \le x \le \pi. \end{cases}$$

Then

$$T(t) = \frac{1}{2\pi} \int_{-\pi}^{\pi} e^{-i\psi(x)} e^{ixt}\, dx = \frac{1}{2\pi i} \left[\frac{1 - e^{-i\pi(t-a)}}{t - a} - \frac{1 - e^{i\pi(t-b)}}{t - b} \right].$$

The coefficient for f with respect to $\{T(t - n)\}$ is

$$c_n = \langle f, T(\cdot - n) \rangle_{PW} = \langle f^{\wedge}, e^{-i\psi(\cdot)} e^{-in\cdot}/\sqrt{2\pi} \rangle_{L^2(-\pi,\pi)}$$

$$= \frac{1}{\sqrt{2\pi}} \int_{-\pi}^{0} f^{\wedge}(x) e^{i(n+a)x}\, dx + \frac{1}{\sqrt{2\pi}} \int_{0}^{\pi} f^{\wedge}(x) e^{i(n+b)x}\, dx.$$

To this let us add and subtract

$$\frac{1}{2\sqrt{2\pi}} \int_{-\pi}^{0} f^{\wedge}(x) e^{i(n+b)x}\, dx + \frac{1}{2\sqrt{2\pi}} \int_{0}^{\pi} f^{\wedge}(x) e^{i(n+a)x}\, dx,$$

and recall the definition of the Hilbert transform $f^{\sim}$ of $f \in L^2$ (Appendix B). After some rearrangement we find

$$c_n = \frac{1}{2i} [f^{\sim}(n + a) - f^{\sim}(n + b)] + \frac{1}{2} [f(n + a) + f(n + b)].$$

Another example of this construction is suggested in Problem 7.1.

7.3 Transformations on the Paley–Wiener space

Some commonly occuring transformations defined on $PW_{\pi w}$ are

Dilation	$\delta_a : f(t) \mapsto f(at), \ a > 0.$
Normalized dilation	$\Delta_a : f(t) \mapsto a^{\frac{1}{2}} f(at), \ a > 0.$
Taking powers	$\pi_n : f(t) \mapsto [f(t)]^n, \ n \in \mathbb{N}.$
Translation	$\tau_h : f(t) \mapsto f(t + h), \ h \in \mathbb{R}.$
Hilbert transformation	$\mathcal{H} : f(t) \mapsto f^{\sim}(t).$
Differentiation	$(\mathcal{D})^{(\beta)} : f(t) \mapsto f^{(\beta)}(t), \ \beta \in \mathbb{N}.$

Some facts about these transformations will be derived in the next few lemmas.

Lemma 7.11 *Normalized dilation is an isometric operator on $L^2(\mathbb{R})$. We have $\Delta_a : PW_{\pi w} \mapsto PW_{aw}$, and if $a < 1$ we have $PW_{aw} \subset PW_{\pi w}$, so in this case Δ_a may be said to leave $PW_{\pi w}$ invariant.*

Proof By direct calculation, and using Appendix A, no. 4.

We need a definition to state our next lemma. Suppose that N and $N^\perp$ are a pair of complementary subspaces of a Hilbert space H; that is, H $= N \oplus N^\perp$.

Definition 7.12 *The pair $\{N, N^\perp\}$ is said to reduce the operator T acting on H if $T : N \mapsto N$ and $T : N^\perp \mapsto N^\perp$.*

Lemma 7.13 *The class of all translations $\{\tau_h\}$, $h \in \mathbb{R}$, together with composition of translations, is a group of unitary operators on $L^2(\mathbb{R})$ which is isomorphic to the additive group $\mathbb{R}$. Each τ_h is reduced by $\{PW_{\pi w}, PW_{\pi w}^\perp\}$.*

Proof Simple calculations show that τ_h is unitary.

Since $\tau_h \tau_{h'} = \tau_{h'+h}$, the isomorphism is given by $\tau_h \leftrightarrow h$, $h \in \mathbb{R}$. As to the last assertion, we note that since supp $f^\wedge \subseteq \{x : |x| \leq \pi w\}$ if $f \in PW_{\pi w}$, and supp $f^\wedge \subseteq \{x : |x| > \pi w\}$ if $f^\wedge \in PW_{\pi w}^\perp$, we need only appeal to the operational formula 3a of Appendix A.

Lemma 7.14 *$-i\mathcal{D}$ is a bounded, linear, self-adjoint operator on $PW_{\pi w}$; it is not invertible there.*

Proof The proof is based on the operational formula (Appendix A no. 6)

$$-i\left(\mathcal{D}f\right)(t) = \left(\mathcal{F}^{-1}xf^\wedge(x)\right)(t).$$

To show that $-i\mathcal{D}$ is bounded, we have

$$\| -i\mathcal{D}f \|^2 = \| -i\mathcal{F}\mathcal{D}f \|^2 = \frac{1}{\sqrt{2\pi}} \int_{-\pi w}^{\pi w} |xf^\wedge(x)|^2 \, dx$$

$$\leq \frac{(\pi w)^2}{\sqrt{2\pi}} \|f^\wedge\|^2 = (\text{constant})\|f\|^2.$$

The self-adjointness follows from

$$\langle -i\mathcal{D}f, g \rangle = \langle -i\mathcal{F}\mathcal{D}f, \mathcal{F}g \rangle = \int_{-\pi w}^{\pi w} xf^\wedge(x)\overline{g^\wedge(x)} \, dx$$

$$= \langle \mathcal{F}f, -i\mathcal{F}\mathcal{D}g \rangle = \langle f, -i\mathcal{D}g \rangle.$$

An example shows that $\mathcal{D}$ is not onto. First we note from Appendix A no. 9 that $\text{sinc}^2 wt/2 \in PW_{\pi w}$. Now

$$p(t) = \int_{-\infty}^{t} \text{sinc}^2 \frac{wv}{2} \, dv$$

is one of its primitives, so that every primitive is of the form $p(t) + $ constant. Furthermore, by a special integral (see Problem 6.11),

$$p(\infty) = \int_{-\infty}^{\infty} \operatorname{sinc}^2 \frac{wv}{2} \, dv = \frac{2}{w}.$$

Since $p(-\infty) = 0$, no primitive of $\operatorname{sinc}^2 wt/2$ can vanish at both $-\infty$ and $+\infty$; therefore there is no member of $PW_{\pi w}$ whose derivative is $\operatorname{sinc}^2 wt/2$, and so $\mathcal{D}$ is not invertible on $PW_{\pi w}$.

More information on these operators can be found in Weston (1949); a much more advanced treatment is in Rochberg (1987).

7.4   Oscillatory properties of Paley–Wiener functions

We continue our study of Paley–Wiener functions by asking the question: *What kind of oscillatory behaviour is imposed on a Paley–Wiener function f by the property of having compactly supported Fourier transform?* For example, does oscillatory behaviour remain as a permanent feature of the function ?

The material in this section relates most naturally to Paley–Wiener functions that are real on the real axis, although this restriction is not always necessary; in fact some of the theorems are general enough not to require it.

What shall we mean by the term "oscillatory behaviour"? In order to arrive at a satisfactory criterion it is useful first to compare Paley–Wiener functions with Gaussians, functions of the form e^{-at^2} (Gaussians are not, of course, Paley–Wiener functions since their Fourier spectrum contains arbitrarily high frequencies (Appendix A, no. 24.)). We do not think of Gaussians as having permanent oscillatory features; they are eventually very smooth, not only because they have derivatives of all orders but also because none of their derivatives has infinitely many zeros. On the other hand, a Paley–Wiener function that is real on the real axis may have infinitely many real zeros; most of the commonly occurring examples do so — the "sinc" function for example, and its powers and translates. For these functions our question is effectively answered; the function oscillates in virtue of its zeros, and the oscillations last for all time since the zeros cannot accumulate. But a Paley–Wiener function need not have infinitely many real zeros; there are examples with only finitely many (Problem 7.6), and even examples with no real zeros at all (Problem 7.7).

It seems then, that our way forward should take account of the derivative, or of higher order derivatives. For example, the Paley–Wiener function $(1-\operatorname{sinc} t)/t^2$ has no real zeros, its first derivative has finitely many, and its second derivative has infinitely many. Let us say that a Paley–Wiener function has *oscillatory behaviour of order N* if its Nth derivative has infinitely many real zeros for some $N \in \mathbb{N}_0$, and N is the least such number. The basic question with which we started would be effectively settled if we could decide whether every Paley–Wiener function has oscillatory behaviour of some order, but it is not known whether this is the case.

We can make one further observation though; as a consequence of Rolle's Theorem, if a Paley–Wiener function f which is real on $\mathbb{R}$ has oscillatory behaviour of order N, then every higher order derivative has infinitely many real zeros. This not true in general if f is complex valued on $\mathbb{R}$, as we can see from Problem 7.5.

The basic fact about zeros of Paley–Wiener functions is the following classical theorem of Titchmarsh, to the effect that every such function has infinitely many zeros in $\mathbb{C}$, and that they have a "sparse" asymptotic distribution.

Theorem 7.15 *Let $f \in PW_{\pi w}$. If $n(r)$ denotes the number of zeros of f in the circle $|z| \leq r$ in the complex z-plane, then*

$$\frac{n(r)}{r} \to 2w, \qquad r \to \infty.$$

When $f \in PW_{\pi w}$ is real on $\mathbb{R}$ and has infinitely many real zeros, the fact that it is band-limited must have an implication about the distribution of the zeros. Approaching this intuitively at first, we could note that since $1/w$ is half a cycle length of the highest frequency component of f, the gaps between consecutive pairs of zeros cannot all be less than $1/w$; if they were, we would have the impossible situation of f oscillating at a frequency higher than its maximum frequency. As a matter of fact, if f is not null the gaps cannot all be equal to $1/w$ either, because then one could legitimately sample f at its zeros and reconstruct by the shifted cardinal series (see the example following Corollary 7.9), which would immediately give a contradiction.

These observations lead us to the striking Theorem 7.16 (below) of Walker (1991a,b), whose proof uses an elegant deployment of two classical inequalities. One of these is *Bernstein's inequality*

$$\|f'\| \leq \pi w \, \|f\|,$$

in which the norm is that of $L^2(\mathbb{R})$. In Theorem 6.7 we saw that this inequality holds in particular for $f \in PW_{\pi w}$.

The other is *Wirtinger's inequality*. Let f be a complex valued function defined on $[a, b]$, with $f \in C^1[a, b]$ and $f(a) = f(b) = 0$. Then

$$\int_a^b |f(x)|^2 \, dx < \left(\frac{b-a}{\pi}\right)^2 \int_a^b |f'(x)|^2 \, dx$$

unless $f(t) = c \sin[(x - a)/(b - a)]$, $c \in \mathbb{C}$.

Wirtinger's inequality is proved for real valued f in Hardy *et al.* (1934, p. 185), and follows for complex valued f by considering real and imaginary parts separately.

Theorem 7.16 *Let $f \in PW_{\pi w}$. Then $|f| > 0$ on at least one open interval of the real axis whose length exceeds $1/w$.*

Proof We need only consider the case in which f has infinitely many zeros on $\mathbb{R}$.

Now suppose that a and b are two consecutive zeros of f, and in order to obtain a contradiction suppose that $b - a \leq 1/w$. Now f is not a sine function on any interval (if it were, its analytic continuation would be a sine and hence not a member of $PW_{\pi w}$), so that Wirtinger's inequality gives

$$\int_a^b |f(x)|^2 \, dx < \left(\frac{b-a}{\pi}\right)^2 \int_a^b |f'(x)|^2 \, dx \le \left(\frac{1}{\pi w}\right)^2 \int_a^b |f'(x)|^2 \, dx.$$

When the left- and right-hand terms are summed over all pairs of consecutive zeros, we obtain

$$\|f\| < \frac{1}{\pi w}\|f'\|,$$

since the zeros cannot accumulate on $\mathbb{R}$. Now combining this with Bernstein's inequality we obtain

$$\|f\| < \frac{1}{\pi w}\|f'\| \le \|f\|,$$

a contradiction.

For example, since wt has only one zero-free interval of length greater than $1/w$, namely $(-1/w, 1/w)$.

It is worth noting that Titchmarsh's theorem does not imply Walker's, because the former gives asymptotic information only, while the latter asserts that there is a "long" gap between zeros somewhere.

One would like to be able to characterize in some convenient way that subclass of $PW_{\pi w}$ whose members have infinitely many zeros on $\mathbb{R}$ (in this context a criterion of Paley and Wiener (1934, p. 75) is relevant). Here we prove the following:

Theorem 7.17 *Let* $f \in PW_{\pi w}$ *and be real valued on* $\mathbb{R}$. *For some* $K \in \mathbb{N}_0$, *let* $f^\wedge$ *have* $K + 1$ *derivatives on* $[-\pi w, \pi w]$, *let* $(f^\wedge)^{(K)}(\pi w) \neq 0$ *whereas* $(f^\wedge)^{(j)}(\pi w) = 0$, $j = 0, 1, \ldots, K - 1$, *and let* $(f^\wedge)^{(K+1)}$ *be absolutely continuous on* $[-\pi w, \pi w]$. *Then* f *has infinitely many real zeros.*

Proof First we show that

$$f(t) = \frac{(-1)^K}{\sqrt{2\pi}(it)^{K+1}}\left[(f^\wedge)^{(K)}(\pi w)e^{i\pi wt} - (f^\wedge)^{(K)}(-\pi w)e^{-i\pi wt}\right] + H_K(t),$$

$$(7.4.1)$$

where

$$H_K(t) = \frac{(-1)^{K+1}}{\sqrt{2\pi}(it)^{K+1}}\int_{-\pi w}^{\pi w} e^{iyt}(f^\wedge)^{(K+1)}(y) \, dy \qquad (7.4.2)$$

and

$$H_K(t) = \mathcal{O}\left(\frac{1}{t^{K+2}}\right). \qquad (7.4.3)$$

To prove these assertions we first integrate by parts in the inverse Fourier transform

$$f(t) = \frac{1}{\sqrt{2\pi}} \int_{-\pi w}^{\pi w} e^{iyt} f^{\wedge}(y)\, dy,$$

to obtain

$$f(t) = -\frac{1}{\sqrt{2\pi} it} \int_{-\pi w}^{\pi w} e^{iyt} (f^{\wedge})'(y)\, dy,$$

since the integrated terms vanish. Then (7.4.1) follows by a further K integrations by parts. Since $(f^{\wedge})^{(K+1)}$ is absolutely continuous we can integrate by parts again in (7.4.2) to obtain

$$H_K(t) = \frac{(-1)^{K+1}}{\sqrt{2\pi}(it)^{K+2}} \left\{ (f^{\wedge})^{(K+1)}(\pi w) e^{i\pi wt} - (f^{\wedge})^{(K+1)}(-\pi w) e^{-i\pi wt} \right\}$$

$$+ \frac{(-1)^{K+2}}{\sqrt{2\pi}(it)^{K+2}} \int_{-\pi w}^{\pi w} e^{iyt} (f^{\wedge})^{(K+2)}(y)\, dy. \tag{7.4.4}$$

This gives (7.4.3).

Since f is real valued we have, from a calculation based on the definition of $\mathcal{F}$, that

$$(-1)^K (f^{\wedge})^{(K)}(-x) = \overline{(f^{\wedge})^{(K)}(x)}.$$

Now with a view to rewriting (7.4.1), we have

$$(f^{\wedge})^{(K)}(\pi w) e^{i\pi wt} - (f^{\wedge})^{(K)}(-\pi w) e^{-i\pi wt}$$

$$= (f^{\wedge})^{(K)}(\pi w) e^{i\pi wt} - (-1)^K (f^{\wedge})^{(K)}(\pi w) e^{-i\pi wt}$$

$$= \begin{cases} 2\,\Re\left[(f^{\wedge})^{(K)}(\pi w) e^{i\pi wt} \right], & n \text{ odd} \\ 2i\,\Im\left[(f^{\wedge})^{(K)}(\pi w) e^{i\pi wt} \right], & n \text{ even.} \end{cases}$$

In either case there exist constants α and β, not both zero, such that from (7.4.1) and (7.4.3)

$$f(t) = \frac{\alpha \cos \pi wt + \beta \sin \pi wt}{t^{K+1}} + \mathcal{O}\left(\frac{1}{t^{K+2}} \right)$$

$$= \frac{1}{t^{K+1}} \left\{ R \sin(\pi wt + \gamma) + \mathcal{O}\left(\frac{1}{t} \right) \right\},$$

and so f has infinitely many real zeros.

This theorem, including the case for which f is complex valued on the real axis, has been treated in detail by Walker (1995).

A Paley–Wiener function is usually called a *band-pass function* if there is a gap, consisting of a single interval centered at the origin, in its spectral support set $[-\pi w, \pi w]$ (more details about band-pass functions can be found in §13.6). While these functions have been extensively studied, little seems to be known about their oscillatory properties. We only have the following lemma, which shows that there can be no example of a band-pass function that belongs to $L^1(\mathbb{R})$, is real valued on $\mathbb{R}$ and fails to have zeros there, unlike the situation in $PW_{\pi w}$ for which Problems 7.7 (b) and (c) provide examples. It is not known whether there is an example with finitely many real zeros.

Lemma 7.18 [2] *A non-null band-pass function that belongs to $L^1(\mathbb{R})$ and is real valued on $\mathbb{R}$ has at least one zero there.*

Proof We can take $f^\wedge$ to be identically zero in an interval containing the origin, so that $f^\wedge(0) = 0$. From the Fourier transform we have, therefore, that $\int_{\mathbb{R}} f(t)\, dt = 0$. Since f is continuous it must change sign somewhere on $\mathbb{R}$.

PROBLEMS

7.1 Investigate the special case $a = -b \neq 0$ of Example 7.10.

7.2 In Corollary 7.9 find the consequences of putting:
    (a) $\psi(x) = x^2$;
    (b) $\omega(x) = |x|^{-\alpha}$, $0 < |\alpha| < \frac{1}{2}$ (this question is quite difficult; a start can be made by consulting Singer (1970, p. 351)).

7.3 $\delta_{1/n}\pi_n : PW_{\pi w} \to PW_{\pi w}$. Is this operator isometric?

7.4 Let $\mathcal{H}$ denote the Hilbert transform. Show that $\mathcal{H}$ maps $PW_{\pi w}$ *onto* itself.

7.5 Show that $(e^{it} - 1)/t$ is a Paley–Wiener function that has infinitely many real zeros, but whose first derivative does not.

7.6 Show that $(1 - \operatorname{sinc} t)/t$ is a Paley–Wiener function with just one real zero.

7.7 Show that the following are Paley–Wiener functions which have no real zeros:

    (a) $-\displaystyle\int_{-\infty}^{t} \frac{\sin^4(\pi x/4)}{(\pi x/4)^3}\, dx$;

    (b) $\operatorname{sinc}^2(t) + \operatorname{sinc}^2(t - a)$, a a sufficiently small real parameter;

    (c) $(1 - \operatorname{sinc} t)/t^2$.

---

[2] This Lemma and its proof were communicated to the author in conversation by M.M. Dodson.

Why does (c), for example, not satisfy the hypotheses of Theorem 7.17?

7.8 Show that there is a Paley–Wiener function, real on the real axis with infinitely many zeros there, but with an arbitrarily long zero-free gap. Start by considering, for any $n \in \mathbb{N}$, the function

$$\frac{\operatorname{sinc} t}{(1-t)(2-t)\ldots(n-t)},$$

and use Ramanujan's formula (Appendix A, no. 15), coupled with recurrence relations for the Gamma function.

7.9 Show that, for any positive integer K, there is a Paley–Wiener function having oscillatory behaviour of order greater than K. Start by defining

$$s_0(t) := (\sin t)/t,$$

$$s_1(t) := \frac{1}{t^2}\left[1 - s_0(t)\right],$$

$$s_2(t) := \frac{1}{t^2}\left[\frac{1}{3!} - s_1(t)\right],$$

$$\vdots$$

and recursively,

$$s_n(t) := \frac{1}{t^2}\left[\frac{1}{(2n-1)!} - s_{n-1}(t)\right].$$

Show that if n is even,

$$s_n(t) = \left[\frac{1}{t^2(2n-1)!} - \frac{1}{t^4(2n-3)!} + \cdots - \frac{1}{t^{2n}} + \frac{\sin t}{t^{2n+1}}\right].$$

Show that $s_n(t)$ has a derivative of order $2n$ with infinitely many zeros, but that its derivative of order $2n-1$ has only finitely many.

8

KRAMER'S LEMMA

In §7.2 we remarked that Hardy's proof of the ordinary cardinal series representation for Paley–Wiener functions exploited the unitary character of the Fourier transform, an integral transform whose inverse has the kernel $e^{ixt}/\sqrt{2\pi}$. It is noticeable that when this kernel is evaluated at $t = n$, the orthonormal basis $\{e^{inx}/\sqrt{2\pi}\}$, $n \in \mathbb{Z}$, for $L^2(-\pi, \pi)$ results. It was recognized long ago (by J.M. Whittaker, 1935 p. 71, for example) that there are other integral transforms with kernels having similar properties, namely that when one argument of the kernel is evaluated at a certain sequence of points the resulting function set is orthogonal over a certain interval; and furthermore, that there is a sampling theorem analogous to the cardinal series. Whittaker suggested the kernel $(xt)^{\frac{1}{2}} J_0(xt)$ of the Bessel–Hankel (or sometimes just Hankel) transform, of order zero. However, no explicit form for the sampling series, or proof, was given.

The situation was formalized by Kramer (1957, p. 68) following an idea of P. Weiss, in a lemma that has been elevated by some writers to the status of theorem in recent years. This is due no doubt to an increased appreciation of its importance, but we shall regard it here as a piece of methodology, adaptable to several different purposes, and continue to use its affectionate name of Kramer's Lemma.

Kramer does give the Bessel case for Bessel functions of general (or perhaps only integer) order, as a special case of a rather formal construction involving kernels arising from self-adjoint differential operators, a topic that will be treated in detail in Chapter 15. For general order ν the Bessel example is discussed in detail in Example 8.7 below.

It is a remarkable fact that the seemingly cumbersome construction that is Kramer's Lemma manages to capture the spirit of a large and varied class of sampling theorems. The ideas in Kramer's Lemma are extremely fruitful in finding and proving sampling theorems, and we shall find ample evidence of this in later pages, particularly in Chapter 15. Now to the lemma itself.

8.1 Kramer's Lemma

Theorem 8.1 (Kramer's Lemma) *Let:*
1. $I \subset \mathbb{R}$ be a bounded interval;
2. $K(x,t)$ be a kernel belonging to $L^2(I)$ for each fixed t in a suitable subset D of $\mathbb{R}$;
3. $\mathbb{X} \subseteq \mathbb{Z}$ be an indexing set, and (λ_n), $n \in \mathbb{X}$, be a sequence of points belonging to D;

4. $\{K(x,\lambda_n)\}$, $n \in \mathbb{X}$, *be an orthogonal basis for* $L^2(I)$. *Also, for any* $g \in L^2(I)$
let

$$f(t) = (\mathcal{K}g)(t) = \int_I g(x)K(x,t)\,dx = \langle g, \overline{K(\cdot,t)}\rangle_{L^2(I)}, \qquad t \in D. \qquad (8.1.1)$$

Then we have the sampling series

$$f(t) = \lim_{N\to\infty}\sum\nolimits_N f(\lambda_n)S_n(t) := \lim_{N\to\infty}\sum_{\substack{|n|\leq N \\ n\in\mathbb{X}}} f(\lambda_n)S_n(t), \qquad (8.1.2)$$

where the "reconstruction functions" S_n *are given by*

$$S_n(t) = \nu_n^{-1}\int_I \overline{K(x,\lambda_n)}K(x,t)\,dx = \nu_n^{-1}\langle \overline{K(\cdot,\lambda_n)}, \overline{K(\cdot,t)}\rangle,$$

and the normalizing factors ν_n *are given by*

$$\nu_n = \int_I |K(x,\lambda_n)|^2\,dx = \|K(\cdot,\lambda_n)\|^2.$$

The sampling series (8.1.2) converges absolutely, and it converges uniformly
over any set $C \subseteq D$ for which $\|K(\cdot,t)\|$ is bounded.

(Let us note carefully the partial sum notation defined in (8.1.2).)

Proof For any positive integer N,

$$\left| f(t) - \sum\nolimits_N f(\lambda_n)S_n(t)\right|$$
$$= \left|\langle g, \overline{K(\cdot,t)}\rangle - \sum\nolimits_N f(\lambda_n)\nu_n^{-1}\langle \overline{K(\cdot,\lambda_n)}, \overline{K(\cdot,t)}\rangle\right|$$
$$= \left|\langle g - \sum\nolimits_N f(\lambda_n)\nu_n^{-1}\overline{K(\cdot,\lambda_n)}, \overline{K(\cdot,t)}\rangle\right|$$
$$\leq \left\|g - \sum\nolimits_N f(\lambda_n)\nu_n^{-1}\overline{K(\cdot,\lambda_n)}\right\| \|K(\cdot,t)\|, \qquad (8.1.3)$$

using Schwarz' inequality for the inner product. The first norm in (8.1.3) is van-
ishingly small for large N, since for each $n \in \mathbb{X}$, $f(\lambda_n)\nu_n^{-1/2} = \langle g, \nu_n^{-1/2}\overline{K(\cdot,\lambda_n)}\rangle$
is the nth Fourier coefficient of $g \in L^2(I)$ with respect to the orthonormal basis
$\{\nu_n^{-1/2}\overline{K(\cdot,\lambda_n)}\}$. So we obtain pointwise converegence for each fixed $t \in D$. If
$\|K(\cdot,t)\|$ is bounded for $t \in C \subseteq D$, then clearly the convergence is uniform on
C.

As a matter of fact there is an even shorter proof which should appeal to those
who like minimalism, and it leads quickly to the absolute convergence. It consists
in nothing more than noticing that, since $\{\nu_n^{-1/2}\overline{K(\cdot,\lambda_n)}\}$ is an orthonormal basis
for $L^2(I)$, the General Parseval Relation (3.1.1) gives

$$\langle g, \overline{K(\cdot, t)} \rangle = \sum_{n \in \mathbf{X}} \left\langle g, \nu_n^{-1/2} \overline{K(\cdot, \lambda_n)} \right\rangle \left\langle K(\cdot, t), \nu_n^{-1/2} K(\cdot, \lambda_n) \right\rangle, \qquad (8.1.4)$$

for every $t \in D$. Now if these inner products are interpreted for the quantities appearing in Kramer's Lemma, then this series *is* the Kramer sampling series.

As to the absolute convergence we have, by Schwarz' inequality for sums,

$$\left\{ \sum_N |f(\lambda_n) S_n(t)| \right\}^2 \leq \sum_N \nu_n^{-1} |f(\lambda_n)|^2 \sum_N \nu_n |S_n(t)|^2.$$

But when $N \to \infty$ each series is convergent, because $\{\nu_n^{-1/2} f(\lambda_n)\}$ are the Fourier coefficients of g (as we have just seen), and by definition $\{\nu_n^{1/2} \overline{S_n(t)}\}$ are the Fourier coefficients of $\overline{K(\cdot, t)}$, both with respect to the orthonormal basis $\{\nu_n^{-1/2} \overline{K(\cdot, \lambda_n)}\}$, and the Parseval Relation (3.1.1) applies.

$$* \qquad * \qquad *$$

There is a result of the Riesz–Fischer type associated with Kramer's Lemma; it is an analogue of, and the proof depends upon, the ordinary Riesz–Fischer theorem in Hilbert space (Theorem 3.3).

Corollary 8.2 *Let the hypotheses of Kramer's Lemma hold, and let* (a_n), $n \in \mathbf{X}$, *be a sequence of complex numbers such that*

$$\sum_{n \in \mathbf{X}} \nu_n^{-1} |a_n|^2 < \infty. \qquad (8.1.5)$$

Then there is a $g \in L^2(I)$ *such that*

$$f(t) := (\mathcal{K}g)(t) = \sum_{n \in \mathbf{X}} a_n S_n(t), \qquad t \in D.$$

Proof From the Riesz–Fischer Theorem (§3.1) and (8.1.5) there is a $g \in L^2(I)$ such that

$$\left\| g - \sum_N \nu_n^{-1/2} a_n \nu_n^{-1/2} \overline{K(\cdot, \lambda_n)} \right\|$$

is vanishingly small for large N. Now

$$\left| f(t) - \sum_N a_n S_n(t) \right|$$

$$= \left| \langle g, \overline{K(\cdot, t)} \rangle - \sum_N a_n \nu_n^{-1} \langle \overline{K(\cdot, \lambda_n)}, \overline{K(\cdot, t)} \rangle \right|$$

$$= \left| \langle g - \sum_N a_n \nu_n^{-1} \overline{K(\cdot, \lambda_n)}, \overline{K(\cdot, t)} \rangle \right|$$

$$\leq \left\| g - \sum_N a_n \nu_n^{-1} \overline{K(\cdot, \lambda_n)} \right\| \, \| \overline{K(\cdot, t)} \|$$

and this is vanishingly small for large N, for every $t \in D$.

Before looking at a few simple examples let us note that the ordinary cardinal series representation for functions belonging to the Paley–Wiener space $PW_{\pi w}$ can be obtained from Kramer's Lemma by taking $K(x,t) = e^{ixt}$ and using the fact that $\{e^{i x n/w}\}$, $n \in \mathbb{Z}$, is an orthogonal basis for $L^2(-\pi w, \pi w)$.

Example 8.3 Take $K(x,t)$ to be $e^{ixt} e^{-ix\alpha}$, where α is any real number. Because $\{e^{ixn/w} e^{-ix\alpha}\}$, $n \in \mathbb{Z}$, is an orthogonal basis for $L^2(-\pi w, \pi w)$, the Kramer method leads to the "shifted" form of the cardinal series,

$$f(t) = \sum_{n \in \mathbb{Z}} f\left(\frac{n}{w} - \alpha\right) \operatorname{sinc} w \left(t - \frac{n}{w} + \alpha\right), \qquad (8.1.6)$$

as originally introduced by E.T. Whittaker (1915). Again, every member of $PW_{\pi w}$ is represented by such a series; that is to say, every f of the form

$$f(t) = \int_{-\pi w}^{\pi w} g(x) e^{ixt}\, dx, \qquad g \in L^2. \qquad (8.1.7)$$

Example 8.4 As a special case of Whittaker's shifted cardinal series let us take $\alpha = -\frac{1}{2}$ and $w = 1$, and in (8.1.1) choose $g(x) = J_0(\sqrt{a^2 - x^2})$. We can now use the special transform (Appendix A, no. 20) in which we choose $a = \pi$, $\nu = 0$, $b = \pi\beta$ and recall that $J_{1/2}(z) = (2/\pi z)^{1/2} \sin z$, and also that the Fourier cosine transform is self-inverse, to write

$$f(t) = \int_0^\pi J_0(\beta\sqrt{a^2 - x^2}) \cos xt\, dx = \frac{\sin(\pi\sqrt{\beta^2 + t^2})}{\sqrt{\beta^2 + t^2}},$$

and substitution into the series (8.1.6) gives

$$\operatorname{sinc} \sqrt{\beta^2 + t^2} = \sum_{n \in \mathbb{Z}} \operatorname{sinc} \sqrt{\beta^2 + (n + \tfrac{1}{2})^2} \, \operatorname{sinc}\left(t - n - \tfrac{1}{2}\right), \qquad (8.1.8)$$

for $\beta, t \in \mathbb{R}$.

There is some interesting history attatched to this last example. When $t = 0$ one can replace $\sum_{n \in \mathbb{Z}}$ in (8.1.8) with $2\sum_{n \in \mathbb{N}_0}$ and obtain the formula

$$\sin \pi\theta = 2\theta \sum_{n \in \mathbb{N}_0} \frac{(-1)^n}{\pi(n + \frac{1}{2})} \frac{\sin \pi \sqrt{\theta^2 + (n + \frac{1}{2})^2}}{\sqrt{\theta^2 + (n + \frac{1}{2})^2}}.$$

Series of this kind were found by Gosper, using symbolic computation. Later, Ismail and Zhang verified the formulae mathematically and generalized some of them, but at this stage the formulae were still surrounded by an aura of mystery.

Then Zayed (1993a) removed all the mystery by noticing that these formulae are just special cases of sampling series. Further series of this type will be found in Problems 8.1 and 8.2.

*   *   *

It follows from the hypotheses of Kramer's Lemma that the integral operator $\mathcal{K}$ is one-to-one, because if $\mathcal{K}g \equiv 0$, then g is orthogonal to every $K(\cdot, \lambda_n)$ and is therefore null.

By assuming more about the transform $\mathcal{K}$ one can give Kramer's Lemma added precision. In the two corollaries to follow we are going to assume that $\mathcal{K}$ maps $L^2(I)$ into $L^2(D)$. Let us denote the range of $\mathcal{K}$ by $\mathfrak{K}$.

Corollary 8.5 *If $\mathcal{K}^{-1}$ is bounded, then $\mathfrak{K}$ is a Hilbert space with reproducing kernel.*

Proof The boundedness of $\mathcal{K}^{-1}$ implies that $\mathfrak{K}$ is a closed and hence complete subspace of $L^2(D)$.

For each $f \in \mathfrak{K}$ there exists $g \in L^2(I)$ such that

$$f(t) = \int_I K(x,t)g(x)\,dx$$
$$= \int_I K(x,t)\left(\mathcal{K}^{-1}f\right)(x)\,dx.$$

By Schwarz' inequality,

$$|l_t f| := |f(t)| \le \|K(\cdot,t)\|\,\|\mathcal{K}^{-1}f\|$$
$$\le \|K(\cdot,t)\|\,\kappa\|f\|,$$

for some constant κ; hence "point evaluation" is a bounded linear functional on $\mathfrak{K}$ and so, by Property 1 of §3.4, $\mathfrak{K}$ has reproducing kernel.

Corollary 8.6 *Suppose that $\mathfrak{K}$ is a Hilbert space with reproducing kernel $k(x,t)$. Then:*
1. *If $\{\varphi_n\}$ is any orthonormal basis for $\mathfrak{K}$, the expansion in $\{\varphi_n\}$ for $f \in L^2(D)$ converges pointwise to $\langle f, k(\cdot,t)\rangle$, the orthogonal projection of f on $\mathfrak{K}$.*
2. *If $\mathcal{K}$ is an isometry, then*

$$k(x,t) = \langle K(x,\cdot), K(y,\cdot)\rangle_{L^2(I)}.$$

Proof
1. Throughout this part of the proof let $\langle,\rangle$ denote the inner product in $L^2(D)$. Put $a_n = \langle f, \varphi_n\rangle$. Then

$$\left| \sum_N a_n \varphi_n(t) - \langle f, k(\cdot, t) \rangle \right|$$
$$= \left| \langle f, \sum_N \varphi_n(\cdot) \overline{\varphi_n(t)} - k(\cdot, t) \rangle \right|$$
$$\leq \|f\| \left\| \sum_N \varphi_n(\cdot) \overline{\varphi_n(t)} - k(\cdot, t) \right\|$$

and the second norm is vanishingly small for large N, by Property 3 of §3.4.

2. If $f \in \mathfrak{R}$, then

$$f = \int_I u(x) K(x, t)\, dx, \qquad u \in L^2(I),$$
$$= \langle K\overline{K}, f \rangle$$

so that, by uniqueness of k,

$$k(x, t) = \int_I K(x, v) \overline{K}(t, v)\, dv$$

as required.

Example 8.7 This is the "Bessel" case. A brief description only will be given; details can be found in Higgins (1972). Take $K(x,t) = (xt)^{1/2} J_\nu(xt)$ in Kramer's Lemma; then the kernel is that of the Bessel–Hankel transform. Now take λ_n to be the nth positive zero of $J_\nu(t)$. Then $\{x^{1/2} J_\nu(x\lambda_n)\}$ is a *Fourier–Bessel set*, known to form an orthogonal basis for $L^2(0,1)$. This means that Kramer's Lemma applies. It is well known that the Bessel–Hankel transform of order ν, $\nu > -1$, is a bounded (indeed unitary), linear operator on $L^2(\mathbb{R}^+)$, and is self-inverse. Hence, Corollary 8.5 applies, and the image of $L^2(0,1)$ by the Bessel–Hankel transform is a reproducing kernel Hilbert subspace of $L^2(\mathbb{R}^+)$; it is the "Bessel analogue" of the Paley–Wiener space. It can be shown, by using special function formulae, that the reproducing kernel is

$$\frac{(xt)^{\frac{1}{2}}}{t^2 - x^2} \{ t J_{\nu+1}(t) J_\nu(x) - x J_{\nu+1}(x) J_\nu(t) \},$$

and that

$$f(t) = \sum_{\lambda_n > 0} f(\lambda_n) \frac{2(\lambda_n t)^{\frac{1}{2}} J_\nu(t)}{J_\nu'(\lambda_n)(t^2 - \lambda_n^2)}$$

for every f of the form

$$f(t) = \int_0^1 g(x)(xt)^{\frac{1}{2}} J_\nu(xt)\, dx, \qquad g \in L^2(0,1). \tag{8.1.9}$$

The following biorthogonal formulation of Kramer's Lemma is useful, particularly for application to non-self-adjoint eigenvalue problems, such as Example 15.13.

Theorem 8.8 *Let:*
1. *$I \subset \mathbb{R}$ be a bounded interval;*
2. *$K(x,t), K^{\star}(x,t)$ be kernels both belonging to $L^2(I)$ for each fixed t in a suitable subset D of $\mathbb{R}$;*
3. *$\{\lambda_n\}, \{\lambda_n^{\star}\}$, $n \in \mathbb{Z}$, be complex sequences, and put $\varphi_n(x) = K(x,\lambda_n)$, $\varphi_n^{\star}(x) = K^{\star}(x,\lambda_n^{\star})$;*
4. *$\{\varphi_n, \varphi_n^{\star}\}$ be a biorthogonal system such that $\{\varphi_n\}$ (and hence $\{\varphi_n^{\star}\}$) is a basis for $L^2(I)$.*
Then if, for some $g \in L^2(I)$,

$$f(t) = \int_I g(x) K(x,t)\, dx, \qquad t \in D,$$

we have

$$f(t) = \lim_{N \to \infty} \sum_N f(\lambda_n) S_n(t), \qquad (8.1.10)$$

where

$$S_n(t) = \nu_n^{-1} \int_I \overline{\varphi_n^{\star}}(x) K(x,t)\, dx \qquad and \qquad \nu_n = \int_I \varphi_n(x) \overline{\varphi_n^{\star}}(x)\, dx.$$

Dually, if for some $\gamma \in L^2(I)$,

$$f(t) = \int_I \gamma(x) K^{\star}(x,t)\, dx$$

we have

$$f(t) = \lim_{N \to \infty} \sum_N f(\lambda_n^{\star}) S_n^{\star}(t), \qquad (8.1.11)$$

where

$$S_n^{\star}(t) = (\overline{\nu_n})^{-1} \int_I \overline{\varphi_n}(x) K^{\star}(x,t)\, dx.$$

The series (8.1.10) converges uniformly on any set $C \subseteq D$ for which $\|K(\cdot,t)\|$ is bounded; and similarly for the dual series (8.1.11) with respect to $\|K^{\star}(\cdot,t)\|$.

The proof is almost identical to the proof of Kramer's Lemma, except that here g is given its biorthogonal expansion in the set $\{\sqrt{K^{\star}(x,\lambda_n^{\star})}\}$ for the first part, and for the dual part γ is given its biorthogonal expansion in the set $\{\sqrt{K(x,\lambda_n)}\}$.

8.2   The Walsh sampling theorem

A sampling theorem in the context of dyadic analysis was proved by Kak by means of Kramer's Lemma. This theorem, often called the Walsh sampling theorem, involves the notion of *sequency*, the analogue in dyadic analysis of frequency in ordinary Fourier analysis. The unit of sequency, using one of its standard definitions, is one half the number of zero crossings of a function per second. The Walsh sampling theorem states, roughly, that if a function $f(t)$ has its sequency content confined to the interval $[0, 2^k)$ for some positive integer k, then

$$f(t) = \sum_{n \in \mathbb{Z}} f(n/2^k) \chi_{[n/2^k,(n+1)/2^k]}(t).$$

The theorem had been obtained originally by Pichler using a different method, and he pointed out that it is effectively a trivial result because it asserts that a "sequency-limited" function must be a step function. The triviality arises because this fact is known *a priori*. It arises in practice because the sequency-limiting operation is effected by decomposing the domain of the function into intervals and replacing the function on each interval by its average over that interval; this of course produces a step function. A direct mathematical proof of this step-function property was given by Splettstößer (1978b).

Because of its triviality this result is not of direct interest to us, and will not be discussed further here. Its main mathematical interest is that it furnishes an example of the very general sampling theorem of Kluvánek (1965); this topic is planned for inclusion in H–S. The interested reader should consult the articles by Pichler (1973), Splettstößer (1978) and Butzer *et al.* (1988, §6.4) for further information and references.

Several other adaptations of Kramer's Lemma are possible. For example, there is a multi-dimensional version in Zayed (1993b, Ch. 8); another possibility is indicated in Problem 8.3 below. The Lemma has been used to prove several sampling theorems associated with special functions, for example the Legendre, associated Legendre, Gegenbauer, Chebyschev, prolate spheroidal, Laguerre and Hermite functions (further information and references are in Zayed, 1991).

PROBLEMS

8.1 Let

$$J(\nu, b, t) := \frac{J_{\nu+1/2}(\sqrt{b^2 + t^2})}{(b^2 + t^2)^{(\nu+1/2)/2}}.$$

Now choose $g(x)$ in (8.1.7) to be $(1 - x^2)^{\nu/2} J_\nu(b\sqrt{1 - x^2})$. Then, using the transform of Appendix A, no. 20 with $a = 1$, show that (8.1.7) becomes

$$f(t) = \int_0^1 (1 - x^2)^{\nu/2} J_\nu(b\sqrt{1-x^2}) \cos xt \, dx = \sqrt{\frac{\pi}{2}} b^\nu J(\nu, b, t), \qquad \nu > -1,$$

and that (8.1.6) becomes

$$J(\nu, b, t) = \sum_{n \in \mathbb{Z}} J(\nu, b, t_n) \operatorname{sinc}(t - t_n),$$

where $t_n = n + \frac{1}{2}$. Show that a special case of this is

$$\frac{J_\nu(b)}{b^\nu} = \frac{2}{\pi} \sum_{n \in \mathbb{N}_0} \frac{(-1)^n}{n + \frac{1}{2}} J(\nu - \tfrac{1}{2}, b, t_n), \qquad \nu > -\tfrac{1}{2}.$$

By letting $b \to 0$, show that a further special case is

$$\left(\frac{\pi}{4}\right)^{\nu+1} \frac{1}{\Gamma(\nu+1)} = \sum_{n \in \mathbb{N}_0} \frac{(-1)^n J_\nu(\pi[n + \frac{1}{2}])}{(2n+1)^{\nu+1}},$$

and that, in turn, a special case of this is the well known formula

$$\sum_{n \in \mathbb{N}_0} \frac{1}{(2n+1)^2} = \frac{\pi^2}{8}$$

(Zayed 1993b, p. 193).

8.2 Show that

$$J_{\nu+t}(x) J_{\nu-t}(x) = \sum_{n \in \mathbb{Z}} J_{\nu+n+1/2}(x) J_{\nu-n-1/2}(x) \operatorname{sinc}(t - n - \tfrac{1}{2}),$$

and that

$$\{J_\nu(x)\}^2 = \frac{4}{\pi} \sum_{n \in \mathbb{N}_0} J_{\nu+n+1/2}(x) J_{\nu-n-1/2}(x) \frac{(-1)^n}{2n+1}.$$

Show that a special case is

$$\sin^2 x = 2x \sum_{n \in \mathbb{N}_0} J_{n+1}(x) J_n(x) \frac{1}{2n+1}.$$

A special cosine transform will be needed; see Zayed (1993b, p. 195).

8.3 Obtain a version of the biorthogonal Kramer's Lemma in which the notion of biorthonormal expansion is replaced with one of the "canonical" frame expansions of Theorem 3.18.

8.4 Show that the sample points $\{\lambda_n\}$ form a set of uniqueness for the class $\{f: f = \mathcal{K}u, \ u \in L^2(I)\}$.

8.5 Prove Kramer's Lemma by first replacing $f(\lambda_n)$, rather than $S_n(t)$, with an inner product in the first line of (8.1.3). This is Kramer's original method, and it shows clearly how the expansion of the kernel K in a function set involving K itself causes samples of the object function f to appear in the series.

8.6 Investigate the consequences of putting

$$g(x) = \frac{2^{\nu-\mu+1}}{\Gamma(\mu-\nu)} x^{\nu+1/2}(1-x^2)^{\mu-\nu-1}, \qquad -1 < \nu < \mu,$$

in (8.1.9), and then using a special Bessel–Hankel transform (Oberhettinger, 1972, p. 80, no. 10.2) to obtain

$$f(t) = t^{\nu-\mu+1/2} J_\mu(t).$$

9

CONTOUR INTEGRAL METHODS

Contour integration has a variety of different uses in sampling theory, and three theorems and their proofs have been selected to illustrate this classical technique. The first is the famous theorem of Paley and Wiener, one of sampling theory's cornerstones. In §9.2 we take the first step towards another important principle of the subject, the equivalence of the sampling theorem with other basic theorems. Lastly, Theorem 9.4 has been included because it is a sampling theorem of some generality and variety (although by no means the most general of its kind; it has been extensively generalized by Hinsen (1993), whose methods are much more refined than those given here). Contour integration provides a powerful method for generating the correct form of the series, as a sum of residues, and of demonstrating the convergence in the symmetric partial sum mode.

9.1 The Paley–Wiener theorem

Paley–Wiener theorems were reviewed in §7.1. Here we choose the original theorem of Paley and Wiener (1934, p. 12) as our first illustration of the contour integral method.

Theorem 9.1 (Paley–Wiener Theorem) *Let $f \in L^2(\mathbb{R})$. Then f has an analytic extension to $\mathbb{C}$ which belongs to E_σ if and only if supp $f^\wedge \subseteq [-\sigma, \sigma]$.*

Proof The simpler half requires us to show that $f \in E_\sigma$ if supp $f^\wedge \subseteq [-\sigma, \sigma]$. Now

$$f(t) = \frac{1}{\sqrt{2\pi}} \int_{\mathbb{R}} f^\wedge(x) e^{-ixt}\, dx \qquad (9.1.1)$$

in the norm of $L^2(\mathbb{R})$, but under the condition on supp $f^\wedge$ this reduces to

$$f(t) = \frac{1}{\sqrt{2\pi}} \int_{-\sigma}^{\sigma} f^\wedge(x) e^{-ixt}\, dx,$$

which holds pointwise for $t \in \mathbb{R}$, by (2.15). This formula gives the analytic continuation of f to all of $\mathbb{C}$; furthermore, a simple estimate gives

$$|f(t)| \le \frac{1}{\sqrt{2\pi}} \|f^\wedge\|_1 e^{\sigma|t|}, \qquad t \in \mathbb{C},$$

so that $f \in E_\sigma$.

The real substance of the Paley–Wiener theorem is in the other half, where we assume that $f \in E_\sigma$ and it must be shown that $\operatorname{supp} f^\wedge \subseteq [-\sigma, \sigma]$. For this purpose we consider the contour integral

$$I_N = \int_{C_N} f(z)e^{-ix\zeta}dz, \qquad N \in \mathbb{N},$$

where C_N is a rectangular contour in the upper half of $\mathbb{C}$ with corners at $z = \pm N$, $(\pm 1 + i)N$. Then

$$I_N = \int_0^N f(N + iv)e^{-ix(N+iv)}i\,dv + \int_N^{-N} f(u + iN)e^{-ix(u+iN)}\,du$$
$$+ \int_N^0 f(-N + iv)e^{-ix(-N+iv)}i\,dv + \int_{-N}^N f(u)e^{-ixu}\,du = 0, \qquad (9.1.2)$$

by Cauchy's theorem.

In the norm of $L^2(\mathbb{R})$ the last integral tends to $f^\wedge$ (apart from a multiplicative constant) as $N \to \infty$. By showing that the first three integrals tend to zero as $N \to \infty$ whenever $|x| > \sigma$ it will follow that the last integral will do the same; this will show that $f^\wedge$ is null outside $[-\sigma, \sigma]$ and our proof will be complete.

The proof requires the following two auxiliary results, (A) and (B); we shall need (B) again in the next section.

(A)   If $f \in E_\sigma$ and $f(u) \to 0$ as $|u| \to \infty$, then $f(u + iv) \to 0$ as $|u| \to \infty$ uniformly in any horizontal strip.

(B)   If f is entire and satisfies $|f(\zeta)| \le \alpha e^{\beta|\zeta|}$, $\zeta \in \mathbb{C}$, and $|f(u)| \le M$, $u \in \mathbb{R}$, then

$$|f(u + iv)| \le Me^{\beta|v|}.$$

These auxiliary results can be found in Young (1980, pp. 82–83).

Let $x < -\sigma$. Then for the first integral, I_1 say, in (9.1.2),

$$|I_1| \le \left(\int_0^R + \int_R^N \right) e^{xv}|f(N + iv)|\,dv$$

for every fixed positive $R < N$. Now from the Plancherel–Pólya inequality (Theorem 6.10) it follows that for any real increasing sequence (λ_n) such that $\lambda_{n+1} - \lambda_n \ge \delta > 0$, $f(\lambda_n) \to 0$ as $|n| \to \infty$; hence $f(t) \to 0$ as $|t| \to \infty$. In particular, f is bounded over $\mathbb{R}$, say $|f(t)| \le M$.

Our auxiliary result (A) above now shows that $f(N + iv) \to 0$ as $N \to \infty$ uniformly in v for $0 < v < R$. Hence

$$\int_0^R e^{xv}|f(N + iv)|\,dv \to 0 \quad \text{as } N \to \infty.$$

Now, by auxiliary result (B) above,

$$|f(N + iv)| \le Me^{\sigma v}.$$

Hence

$$\int_R^N e^{xv} |f(N + iv)| \, dv \le M \int_R^N e^{(x+\sigma)v} \, dv = \frac{M}{x+\sigma} \left\{ e^{(x+\sigma)N} - e^{(x+\sigma)R} \right\}.$$

Since $x + \sigma < 0$, this can be made arbitrarily small by fixing R at a large value and then leting $N \to \infty$. Hence $I_1 \to 0$ as $N \to \infty$, and the same is true for the third integral in (9.1.2) by a similar argument.

We now consider the second integral in (9.1.2), I_2 say. We have

$$|I_2| \le M \int_{-N}^N e^{\sigma N} e^{Nx} \, du = M \, e^{(x+\sigma)N} \, 2N \to 0 \quad \text{as } n \to \infty,$$

again because $x + \sigma < 0$. The fourth integral in (9.1.2) is handled similarly.

This completes the proof for $x < -\sigma$. The case $x > \sigma$ is handled in a similar way by using a contour in the lower half-plane.

9.2   Some formulae of analysis and their equivalence

We shall show that the sampling theorem belongs to certain groupings of formulae, all members of which are equivalent in the sense that each can be deduced from any of the others by elementary means. We shall present two such groupings. The first involves the sampling theorem, Cauchy's integral formula and Poisson's summation formula; it is the more elementary one, and the functions to which the formulae apply are taken from Bernstein classes. The second involves the sampling theorem and Poisson's formula with more general hypotheses, as well as the summation formulae of Euler–MacLaurin and Abel–Plana.[1] Here, the proofs are beyond our present scope. It should also be mentioned that the formulae hold under a variety of hypotheses, many of them much more general than those mentioned here, our purpose being to show *some* hypotheses under which one can obtain equivalence.

The first grouping consists of the following three theorems.

Theorem A. (Sampling Theorem) *Let $f \in B_{\pi w}^1$, $(w > 0)$. Then*

$$f(z) = \sum_{n \in \mathbb{Z}} f\left(\frac{n}{w}\right) \operatorname{sinc}(wz - n), \tag{9.2.1}$$

---

[1] As the manuscript for this book was nearing completion, Professor P.L. Butzer kindly drew the author's attention to the fact that the celebrated functional equation for the Riemann zeta-function can also be added to this second grouping. The zeta-function is given by $\zeta(z) = \sum_{k=1}^{\infty} k^{-z}$; it is holomorphic for $\Re e\, z > 1$, with a mereomorphic extension to all of $\mathbb{C}$ with just a simple pole at $z = 1$. One of the greatest discoveries of Riemann was that it satisfies the functional equation $\xi(\zeta) = \xi(1 - \zeta)$, where $\xi(\zeta) := (1/2)z(z - 1)\pi^{-z/2}\Gamma(z/2)\zeta(z)$. The reader is referred to Butzer and Nasri-Roudsari (to appear) for further comments, historical remarks and references.

uniformly on compact subsets of $\mathbb{C}$.

It follows by considering the uniform convergence of the differentiated series that we also have

$$\left(\frac{d}{dz}\right)^r f(z) = \sum_{n\in\mathbb{Z}} f\left(\frac{n}{w}\right)\left(\frac{d}{dz}\right)^r \operatorname{sinc}(wz - n), \qquad (9.2.2)$$

for every $r \in \mathbb{N}$.

Theorem B. (Cauchy's Integral Formula — a Special Case) *Let C be a simple, closed, rectifiable, positively oriented curve in* $\mathbb{C}$. *Let $f \in B_\sigma^\infty$ for some $\sigma > 0$. Then, for all $r \in \mathbb{N}_0$,*

$$\left(\frac{d}{dz}\right)^r f(z) = \begin{cases} \dfrac{r!}{2\pi i}\displaystyle\int_C \frac{f(\zeta)}{(\zeta-z)^{r+1}}\,d\zeta, & z \in \operatorname{int} C; \\ 0, & z \in \operatorname{ext} C. \end{cases} \qquad (9.2.3)$$

where int C *denotes the interior of C and* ext C *denotes the exterior of C.*

Theorem C. (Poisson's Summation Formula — a Special Case) *Let $f \in B_{2\pi w}^1$. Then*

$$\sum_{k\in\mathbb{Z}} f\left(\frac{n}{w}\right) = \sqrt{2\pi}\,w f^\wedge(0) = w\int_{\mathbb{R}} f(t)\,dt, \qquad (9.2.4)$$

the series being absolutely convergent.

Theorem 9.2 *Theorem A $\Longleftrightarrow$ Theorem B, and Theorem A $\Longleftrightarrow$ Theorem C.*

Proof First we prove that Theorem B implies Theorem A. We show that (9.2.1) holds in the case $w = 1$; then the general case can be obtained by considering $f(z/w)$. Furthermore, we shall assume for convenience that $f \in B_\sigma^1$, $0 < \sigma < 1$. The case $\sigma = 1$ can subsequently be obtained by considering $\lim_{\delta\to 1^-} f(\delta z)$.

Thus, for $f \in B_\sigma^1$, $0 < \sigma < 1$, let

$$I_m = \frac{\sin \pi z}{2\pi i}\int_{C_m} \frac{f(\zeta)}{(\zeta-z)\sin\pi\zeta}\,d\zeta$$

where C_m is a square contour with corners at $z = (\pm 1 \pm i)(m + \frac{1}{2})$, $m \in \mathbb{N}$; this contour avoids the zeros of $\sin\pi\zeta$. Using the product representation

$$\sin\pi\zeta = \lim_{N\to\infty} s_N = \lim_{N\to\infty} \pi\zeta\prod_{j=1}^N \left(1 - \frac{\zeta^2}{j^2}\right)$$

we can write

$$I_m = \lim_{N\to\infty} \frac{\sin\pi z}{2\pi i}\int_{C_m} \frac{f(\zeta)}{(\zeta-z)s_N(\zeta)}\,d\zeta. \qquad (9.2.5)$$

In the integrand we need the partial fractions decomposition

$$\frac{1}{(\zeta - z)s_N(\zeta)} = \frac{1}{(\zeta - z)s_N(z)} + \sum_{|n|\leq N} \frac{1}{(\zeta - n)(n - z)s_N'(n)}.$$

This is proved using the fact that if $p(\zeta)$ is a polynomial whose only zeros are simple zeros at $r_1, r_2, \ldots r_s$, then

$$\frac{1}{p(\zeta)} = \sum_{j=1}^{s} \frac{1}{(\zeta - r_j)p'(r_j)}.$$

On multiplying by $f(\zeta)$ and using Cauchy's formula, we obtain

$$\frac{1}{2\pi i}\int_{C_m} \frac{f(\zeta)}{(\zeta - z)s_N(\zeta)}\, d\zeta = \frac{f(z)}{s_N(z)} + \sum_{|n|\leq m} \frac{f(n)}{(n - z)s_N'(n)}$$

if $m < N$. Now because $\lim_{N\to\infty} s_N'(n) = \pi(-1)^n$ we have, from (9.2.5),

$$I_m(z) = f(z) - \sin \pi z \sum_{|n|\leq m} f(n)\frac{(-1)^n}{\pi(z - n)}$$

$$= f(z) - \sum_{|n|\leq m} f(n)\operatorname{sinc}(z - n).$$

we need to show that $I_m(z) \to 0$ as $m \to \infty$, uniformly on compact subsets of $\mathbb{C}$. It will suffice, therefore, to prove this if z is any point of a closed disc Δ_R, of radius $R > 0$ and centred at the origin. When $m + \frac{1}{2} > R$ and $\zeta \in C_m$, we have $|\zeta - z| \geq m + \frac{1}{2} - R$, so that

$$|I_m(z)| \leq \frac{|\sin \pi z|}{2\pi(m + \frac{1}{2} - R)} \int_{C_m} \left|\frac{f(\zeta)}{\sin \pi\zeta}\right| d\zeta. \qquad (9.2.6)$$

Let $\zeta = u + iv$. Then, on the side $C_m^{(1)}$ of C_m which joins $(1 - i)(m + \frac{1}{2})$ to $(1 + i)(m + \frac{1}{2})$, we have $\zeta = m + \frac{1}{2} + iv$, so that

$$|\sin \pi\zeta|^2 = \sin^2 \pi u + \sinh^2 \pi v = 1 + \sinh^2 \pi v = \cosh^2 \pi v \geq e^{2\pi|v|}.$$

Throughout the remainder of the proof K will denote a constant that will generally have a different value at each appearance.

We also have $|f(\zeta)| \leq K\, e^{\pi\sigma|v|}$, so that

$$\int_{C_m^{(1)}} \left| \frac{f(\zeta)}{\sin \pi \zeta} \right| d\zeta \leq K \int_{-m-\frac{1}{2}}^{m+\frac{1}{2}} e^{\pi|v|(\sigma-1)} \, dv$$

$$= K \frac{e^{\pi(\sigma-1)(m+\frac{1}{2})} - 1}{\pi(\sigma-1)},$$

which remains bounded as $m \to \infty$. A similar estimate holds on the side $C_m^{(3)}$ of C_m which joins $(-1+i)(m+\frac{1}{2})$ to $(-1-i)(m+\frac{1}{2})$.

On the side $C_m^{(2)}$ of C_m which joins $(1+i)(m+\frac{1}{2})$ to $(-1+i)(m+\frac{1}{2})$ we have $\zeta = u + i(m+\frac{1}{2})$, and $|\sin \pi \zeta| \geq \sinh \pi(m+\frac{1}{2})$. Hence

$$\int_{C_m^{(2)}} \left| \frac{f(\zeta)}{\sin \pi \zeta} \right| d\zeta \leq K \int_{-m-\frac{1}{2}}^{m+\frac{1}{2}} \frac{e^{\pi|v|\sigma}}{\sinh \pi(m+\frac{1}{2})} \, dv$$

$$= K \frac{e^{\pi\sigma(m+\frac{1}{2})} - 1}{\sinh \pi(m+\frac{1}{2})},$$

which vanishes as $m \to \infty$ since $\sigma < 1$. Again, the integral along $C_m^{(4)}$ is handled in the same way. These estimates for the four separate parts of C_m can now be used in conjunction with (9.2.6) to give the required result, and this completes the proof of Theorem B $\implies$ Theorem A.

We now turn to the converse implication. Let Theorem A hold. Then we have (9.2.1), and because of its uniform convergence,

$$\int_C \frac{f(\zeta)}{(\zeta-z)^{r+1}} \, d\zeta = \sum_{n \in \mathbb{Z}} f\left(\frac{n}{w}\right) \int_C \frac{\text{sinc}(w\zeta - k)}{(\zeta - z)^{r+1}} \, d\zeta \qquad (9.2.7)$$

if z does not lie on C. The proof for the case $f \in B_\sigma^1$ will be easily completed if we can show (without using the Cauchy integral theory of course) that

$$\int_C \frac{\text{sinc}(w\zeta - k)}{(\zeta - z)^{r+1}} \, d\zeta = \begin{cases} \dfrac{2\pi i}{r!} \left(\dfrac{d}{dz}\right)^r \text{sinc}(wz - n), & z \in \text{int } C; \\ 0, & z \in \text{ext } C. \end{cases} \qquad (9.2.8)$$

This is because the right-hand side of (9.2.7) will then be $(d/dz)^r f(z)$ when $z \in \text{int } C$, and it will vanish if $z \in \text{ext } C$.

To do this we first expand the sinc function into an appropriate power series, as follows:

$$\text{sinc}(w\zeta - k) = \frac{\sin \pi(w\zeta - k)}{\pi(w\zeta - k)} = \sum_{j=0}^{\infty} \frac{(-1)^j (\pi w)^{2j}}{(2j+1)!} \left(\zeta - \frac{n}{w}\right)^{2j}$$

$$= \sum_{j=0}^{\infty} \frac{(-1)^j (\pi w)^{2j}}{(2j+1)!} \sum_{s=0}^{2j} \binom{2j}{s} \left(z - \frac{n}{w}\right)^{2j-s} (\zeta - z)^s. \qquad (9.2.9)$$

To carry out the integration indicated in (9.2.8) the series (9.2.9) is inserted on the left. Then integration formulae of the following kind will be needed.

$$\int_C (\zeta - z)^{s-r-1}\, d\zeta = \begin{cases} \begin{cases} 2\pi i, & s = r \\ 0, & s \neq r, \end{cases} & z \in \text{int } C \\ \\ 0, & z \in \text{ext } C. \end{cases} \qquad (9.2.10)$$

These can be proved directly; the reader is invited to do this in Problem 9.1. They show that the integral on the left of (9.2.8) vanishes when $z \in \text{ext } C$ as required; and when $z \in \text{int } C$, it is

$$2\pi i \sum_{j=0}^{\infty} \frac{(-1)^j (\pi w)^{2j}}{(2j+1)!} \binom{2j}{r} \left(z - \frac{n}{w}\right)^{2j-r}$$

$$= \frac{2\pi i}{r!} \left(\frac{d}{dz}\right)^r \sum_{j=0}^{\infty} \frac{(-1)^j (\pi w)^{2j}}{(2j+1)!} \left(z - \frac{n}{w}\right)^{2j} = \frac{2\pi i}{r!} \left(\frac{d}{dz}\right)^r \text{sinc}(wz - n),$$

again as required in (9.2.8). This completes the proof of Theorem A $\implies$ Theorem B.

Our next task is to show that Theorem A is equivalent to Theorem C. First let Theorem A hold. Since (9.2.1) converges uniformly over any compact subset of $\mathbb{C}$, we can choose $A > 0$ and integrate as follows:

$$\int_{-A}^{A} f(t)\, dt = \sum_{n \in \mathbb{Z}} f\left(\frac{n}{w}\right) \int_{-A}^{A} \text{sinc}(wt - k)\, dt.$$

Now $f \in L(\mathbb{R})$, so we can let $A \to \infty$. Using the special (improper Riemann) integral $\int_{\mathbb{R}} \text{sinc}(wt - k)\, dt = 1/w$ and the fact that $B_{\pi w}^1 \subset B_{2\pi w}^1$, we obtain Theorem C.

We prove the converse implication in the case $w = 1$; when w is arbitrary a scaling argument can be used as before. Thus, let Theorem C hold, and let (9.2.4) be applied to $f(t)\,\text{sinc}(z-t)$, which belongs to $B_{2\pi}$ because f does. Then we have

$$\sum_{k \in \mathbb{Z}} f(n)\,\text{sinc}(z - n) = \int_{\mathbb{R}} f(u)\,\text{sinc}(z - u)\, du = \frac{1}{\sqrt{2\pi}} \int_{-\pi}^{\pi} f^\wedge(x)e^{ixz}\, dx = f(t),$$

the second equality being a consequence of Plancherel's theorem, since $f \in B_{2\pi}^1 \subset B_{2\pi}^2$, and the third by Fourier inversion.

This completes the proof, and completes the proof of the equivalence of Theorems A, B and C.

There is a further grouping of four equivalent formulae; they are listed below as Theorems A*, C*, D* and E*. The first two are forms of the sampling theorem and Poisson's summation formulae which are more general than Theorems A and C above. To these we can add Theorems D* and E*, which are appropriate formulations of the summation formulae of Euler–MacLaurin and Abel–Plana respectively. There is no appropriate form of Cauchy's theorem in this context.

Theorem A*. (Generalized Sampling Theorem)   *Let $f \in L^1(\mathbb{R}) \cap C(\mathbb{R})$ and let $f^\wedge \in L^1(\mathbb{R})$. Then*

$$f(t) = \sum_{k \in \mathbb{Z}} f\left(\frac{k}{w}\right) \operatorname{sinc}(wt - k)$$

$$+ \sqrt{2\pi} \sum_{k \in \mathbb{Z}} \left(1 - e^{2\pi i k w t}\right) \int_{(2k-1)\pi w}^{(2k+1)\pi w} f^\wedge(x) e^{ixt}\, dx.$$

The second term is recognized as the *error due to aliasing* (see §11.3).

Theorem C*. (Poisson's Summation Formula)   *Let f be an absolutely continuous function on $\mathbb{R}$, and let both f and f' belong to $L^1(\mathbb{R})$. Then*

$$A^{1/2} \sum_{k \in \mathbb{Z}} f(Ak) = B^{1/2} \sum_{k \in \mathbb{Z}} f^\wedge(Bk), \qquad A > 0, \;\; AB = 2\pi.$$

Theorem D*. (Euler–MacLaurin Summation Formula)   *Suppose that $f \in C^{(2r+1)}[n, m]$, where $r \in \mathbb{N}_0$ and $m, n \in \mathbb{Z}$, $n < m$. Then*

$$\sum_{k=n}^{m} f(k) = \int_n^m f(t)\, dt + \tfrac{1}{2}\{f(m) + f(n)\}$$

$$+ \sum_{k=1}^{r} \frac{B_{2k}}{(2k)!}\{f^{(2k-1)}(m) - f^{(2k-1)}(n)\}$$

$$+ \sideset{}{'}\sum_{k=-\infty}^{\infty} \frac{1}{(2\pi i k)^{2r+1}} \int_n^m f^{(2r+1)}(t) e^{-2\pi i k t}\, dt,$$

B_{2k} being the Bernoulli numbers.

Theorem E*. (Abel–Plana Summation Formula)   *Let f be holomorphic in $\{z \in \mathbb{C} : \Re e z \geq 0\}$ and let either the series $\sum_{n=0}^{\infty} f(n)$ or the integral $\int_0^\infty f(u)\, du$ be convergent. Also let*

$$\lim_{v \to \infty} |f(u \pm iv)| e^{-2\pi v} = 0$$

uniformly in u on every finite interval, and let

$$\int_0^\infty |f(u \pm iv)| e^{-2\pi v} \, dv$$

exist for every $u \geq 0$ and tend to 0 when $u \to \infty$. Then

$$\sum_{n=0}^\infty f(n) = \tfrac{1}{2} f(0) + \int_0^\infty f(u) \, du + i \int_0^\infty \frac{f(iv) - f(-iv)}{e^{2\pi v} - 1} \, dv.$$

Remarks and references concerning the equivalence of Theorems A*, C* and D* can be found in Butzer and Nasri-Roudsari (to appear); the Abel–Plana formula was added to this grouping by Rahman and Schmeißer (1994).[2]

9.3   A general sampling theorem

As a final illustration of the contour integral method in sampling theory we prove Theorem 9.4 below, which deals with real sample points $\{\lambda_n\}$ satisfying a condition of the type

$$|\lambda_n - n| \leq L, \qquad L > 0, \quad n \in \mathbb{Z}.$$

Throughout this section L will never exceed $\tfrac{1}{4}$, so of course our sampling set $\{\lambda_n\}$ will be uniformly discrete.

Let H be an entire function with zeros at $\{\lambda_n\}$. The method arises from integrating a function of the form

$$F(\zeta) = \frac{f(\zeta)}{H(\zeta)(\zeta - z)}$$

around a suitable sequence of contours. In fact, let C_m be a square contour with corners at $z = (\pm 1 \pm i)(m + \tfrac{1}{2})$, $m \in \mathbb{N}$, as in the last section. Then each member of $\{\lambda_n\}$ is distant at least $\tfrac{1}{4}$ from every contour C_m.

Here we shall choose H to be a power of the canonical product G taken over the points $\{\lambda_n\}$, defined as follows.

Definition 9.3 *Let $\{\lambda_n\}$, $n \in \mathbb{Z}$, be a real sequence of distinct points, and D a positive constant such that $|\lambda_n - n| \leq D$. With little loss of generality we can take $\lambda_0 = 0$. Let*

$$G(\zeta) := \zeta \prod_{n=1}^\infty \left(1 - \frac{\zeta}{\lambda_n}\right) \left(1 - \frac{\zeta}{\lambda_{-n}}\right).$$

The general term of G is $1 + \mathcal{O}(1/n^2)$ for large $|n|$, so that G is absolutely convergent for every $z \in \mathbb{C}$, and represents an entire function whose only zeros are simple and are located at $\{\lambda_n\}$.

---

[2]In response to a suggestion by Professor P.L. Butzer in a lecture given at the University of Erlangen–Nürnberg in February 1990.

Let us put

$$F_r(\zeta) = \frac{f(\zeta)}{G^{r+1}(\zeta)(z-\zeta)}.$$

Now from Cauchy's residue theorem we obtain

$$\lim_{m\to\infty} \left| \frac{f(z)}{G^{r+1}(z)} - \sum_{|n|\leq m} \{\text{residues at } \zeta = \lambda_n \text{ of } F_r(\zeta)\} \right|$$

$$\leq \lim_{m\to\infty} \frac{1}{2\pi} \int_{C_m} |F_r(\zeta)|\, d\zeta. \tag{9.3.1}$$

A sampling theorem is obtained if we can show that the second limit is zero.

Theorem 9.4 *Let $\{\lambda_n\}$ and D be as in the previous definition, and let $r \in \mathbb{N}_0$. Then for every $f \in B^p_\sigma$ we have*

$$f(z) = G^{r+1}(z) \sum_{n\in\mathbb{Z}} \sum_{\nu=0}^{r} f^{(\nu)}(\lambda_n) \sum_{j=0}^{r-\nu} \frac{1}{\nu!(r-\nu-j)!} \frac{Q_{rn\nu j}}{(z-\lambda_n)^{j+1}}, \tag{9.3.2}$$

where

$$Q_{rn\nu j} = \lim_{\zeta\to\lambda_n} \left\{ \left(\frac{\zeta-\lambda_n}{G(\zeta)} \right)^{r+1} \right\}^{(r-\nu-j)}, \tag{9.3.3}$$

with uniform convergence on compact subsets of $\mathbb{C}$, if
either (1) $\sigma < (r+1)\pi$, $D < 1/4(r+1)$ and $1 \leq p \leq \infty$;
or   (2) $\sigma = (r+1)\pi$, $D < 1/4(r+1)p$ and $1 < p \leq 2$.

To prove this sampling theorem we shall need the estimates in the following two lemmas. Throughout the remainder of this section K will denote a constant, possibly depending on f and p, and not necessarily the same at each occurrence.

Lemma 9.5 *Let $f \in B^p_\sigma$, $(1 < p \leq 2)$. Then*

$$|f(\zeta)| \leq K \frac{e^{\sigma|v|}}{(1+|v|)^{1/p}}, \qquad \zeta = u + iv. \tag{9.3.4}$$

Proof Since $f \in B^p_\sigma$ we know that $f^\wedge \in L^q(\mathbb{R})$, and that

$$f(\zeta) = \frac{1}{\sqrt{2\pi}} \int_{-\sigma}^{\sigma} f^\wedge(x) e^{ix\zeta}\, dx.$$

Then, from Hölder's inequality we have

$$|f(\zeta)| \leq K\|f^\wedge\|_q \left| \frac{e^{\sigma vp} - e^{-\sigma vp}}{vp} \right|^{1/p}, \qquad \zeta = u + iv \in \mathbb{C}.$$

Now the right-hand side is bounded for bounded $|v|$, and we have $|v| \geq \frac{1}{2}(1+|v|)$ for large $|v|$; in any case we have (9.3.4).

Lemma 9.6 *Let G be as in Definition 9.3. Then*

$$|G(\zeta)| \geq K \frac{|v|e^{\pi|v|}}{(1+|\zeta|)^{4D+1}}, \qquad \zeta \in \mathbb{C}; \tag{9.3.5}$$

$$|G(\zeta)| \geq K \frac{e^{\pi|v|}}{(1+|\zeta|)^{4D}}, \qquad \zeta = m + \tfrac{1}{2} + iv, \quad m \in \mathbb{Z}. \tag{9.3.6}$$

The inequality (9.3.5) is due to Levinson and (9.3.6) is due to Seip (see Seip (1987) for both of these estimates). We can now return to the proof of Theorem 9.4.

Proof Let the side of C_m that joins $(m+\frac{1}{2})(1-i)$ to $(m+\frac{1}{2})(1+i)$ be denoted by Γ_1, and let the side joining $(m+\frac{1}{2})(1+i)$ to $(m+\frac{1}{2})(-1+i)$ be denoted by Γ_2. To prove part (1), suppose that $f \in B_\sigma^\infty$. For large m and $\zeta \in \Gamma_1$, we have, using auxiliary result (B) of §9.1 and (9.3.6), and recalling that $\zeta = m + \frac{1}{2} + iv$,

$$|F_r(\zeta)| \leq K \frac{e^{(\sigma-(r+1)\pi)|v|}(1+|\zeta|)^{4(r+1)D}}{|\zeta|-|z|},$$

so that

$$\left| \int_{\Gamma_1} F_r(\zeta)\,d\zeta \right| \leq K \sup_{\zeta \in \Gamma_1} \left\{ \frac{(1+|\zeta|)^{4(r+1)D}}{|\zeta|-|z|} \right\} \int_0^{m+\frac{1}{2}} e^{(\sigma-(r+1)\pi)v}\,dv.$$

This expression approaches zero as $m \to \infty$ if $\sigma < (r+1)\pi$ and $4(r+1)D < 1$. Again, let $\zeta \in \Gamma_2$. Now using auxiliary result (B) of §9.1 and (9.3.5), and recalling that $\zeta = u + i(m+\frac{1}{2})$,

$$\left| \int_{\Gamma_2} F_r(\zeta)\,d\zeta \right| \leq K \frac{e^{(\sigma-(r+1)\pi)(m+\frac{1}{2})}}{(m+\frac{1}{2})^{r+1}} \int_{-m-\frac{1}{2}}^{m+\frac{1}{2}} \frac{(1+|\zeta|)^{(r+1)(4D+1)}}{|\zeta|-|z|}\,du,$$

and this expression approaches zero as $m \to \infty$ if $\sigma < (r+1)\pi$, because the exponential term is dominant.

One handles the other sides in the same way.

As to part (2), suppose that $f \in B_\sigma^p$, $1 < p \leq 2$. On using (9.3.4) and (9.3.6) we obtain, for $\zeta \in \Gamma_1$,

$$\left| \int_{\Gamma_1} F_r(\zeta)\,d\zeta \right| \leq K \sup_{\zeta \in \Gamma_1} \left\{ \frac{(1+|\zeta|)^{4(r+1)D}}{|\zeta|-|z|} \right\} \int_0^{m+\frac{1}{2}} \frac{e^{(\sigma-(r+1)\pi)v}}{(1+v)^{1/p}}\,dv,$$

If we put $\sigma = (r+1)\pi$ and recall that $\zeta = m + \frac{1}{2} + iv$, this integral is $\mathcal{O}(m^{1-1/p})$ for large m, so the whole expression approaches zero as $m \to \infty$ if $4(r+1)Dp < 1$.

Similar considerations apply to Γ_2, where $\zeta = u + i(m + \frac{1}{2})$. We use (9.3.4) and (9.3.5) to obtain

$$\left| \int_{\Gamma_2} F_r(\zeta) \, d\zeta \right| \leq K \int_{-m-\frac{1}{2}}^{m+\frac{1}{2}} \frac{e^{(\sigma - (r+1)\pi)|v|}(1 + |\zeta|)^{(r+1)(4D+1)}}{(1 + |v|)^{1/p}|v|^{r+1}(|\zeta| - |z|)} \, du.$$

When $\sigma = (r + 1)\pi$ this integral does not exceed

$$K \sup_{\zeta \in \Gamma_2} \left\{ \frac{(1 + |\zeta|)^{(r+1)(4D+1)}}{(1 + |v|)^{1/p}|v|^{r+1}(|\zeta| - |z|)} \right\} \int_{-m-\frac{1}{2}}^{m+\frac{1}{2}} du$$
$$\leq K m^{4(r+1)D - 1/p}$$

for large m, and hence approaches zero when $m \to \infty$ if $4(r + 1)Dp < 1$.

Again, the other sides are handled in the same way. This shows that the limit on the right-hand side of (9.3.1) is zero. The summand on the left-hand side is

$$\frac{1}{r!} \lim_{\zeta \to \lambda_n} \left\{ \frac{f(\zeta)}{z - \zeta} \left[\frac{\zeta - \lambda_n}{G^{r+1}(\zeta)} \right]^{r+1} \right\}^{(r)},$$

and after two applications of the Leibnitz rule and some manipulation the form (9.3.2) is obtained, and the proof is complete.

It is interesting to notice two special cases of Theorem 9.4. One of these is when $\lambda_n = n$, $n \in \mathbb{Z}$. In this case (9.3.2) becomes

$$f(z) = \sin^{r+1} \pi z \sum_{n \in \mathbb{Z}} \sum_{\nu=0}^{r} f^{(\nu)}(n) \sum_{j=0}^{r-\nu} \frac{1}{\nu!(r - \nu - j)!} \frac{Q_{rn\nu j}}{(z - n)^{j+1}},$$

where

$$Q_{rn\nu j} = \lim_{\zeta \to n} \left\{ \left(\frac{\zeta - n}{\sin \pi \zeta} \right)^{r+1} \right\}^{(r-\nu-j)}.$$

This form of $Q_{rn\nu j}$ can be calculated in terms of generalized Bernoulli numbers (see, e.g., Jerri 1977, p. 1572) or, alternatively, using central factorial numbers (Butzer and Stens 1993, p. 168).

The other case is when only samples of f itself are available, so that $r = 0$. From (9.3.2) we find that the form of Q needed is

$$Q_{0n00} = \lim_{\zeta \to \lambda_n} \frac{\zeta - \lambda_n}{G(\zeta)} = \frac{1}{G'(\lambda_n)},$$

and so we obtain

$$f(z) = \sum_{n \in \mathbb{Z}} f(\lambda_n) \frac{G(z)}{G'(\lambda_n)(z - \lambda_n)},$$

a series of the Lagrange type.

Theorem 9.4 has been arranged to show how an increase in information (in this case samples of derivatives) at each sample point can be exploited by allowing an increase in bandwidth. Alternatively, one could leave the bandwidth constant and use the additional information to increase the sample point spacing. For example, if samples of the function and of its first r derivatives are available at each sample point, an r-fold inrease in sample point spacing is possible (it should be noted that the *sampling rate* remains constant under either alternative).

PROBLEMS

9.1 Prove the integration formulae (9.2.10) without recourse to the Cauchy integral theory. (One can follow the treatment in Apostol 1957, p. 239 and Hille 1959, p. 162.)

9.2 Show (by re-scaling) that there is an alternative form of Theorem (9.4) for $f \in B_\pi^p$, but with increased sample point spacing. Show that its sampling rate is the Nyquist rate, and that this rate holds also for Theorem (9.4). In particular, for $f \in B_{2\pi}^p$,

$$f(z) = \left\{ \frac{\sin \pi z}{\pi} \right\}^2 \sum_{n \in \mathbb{Z}} \left[\frac{f(n)}{(z-n)^2} + \frac{f'(n)}{z-n} \right];$$

or, alternatively, for $f \in B_\pi^p$,

$$f(z) = \left\{ \frac{\sin \pi z/2}{\pi/2} \right\}^2 \sum_{n \in \mathbb{Z}} \left[\frac{f(2n)}{(z-2n)^2} + \frac{f'(2n)}{z-2n} \right].$$

9.3 Let $\sigma(z)$ denote the Weierstrass sigma function, with zeros at the "lattice points" $\{m+in\}$, $(m,n) \in \mathbb{Z}^2$ (properties of $\sigma(z)$ can be found in, for example, Whittaker and Watson 1962, pp. 447–448). Let f be an entire function satisfying the condition

$$\limsup_{r \to \infty} \frac{\log M(r)}{r^2} < \frac{\pi}{2},$$

where $M(r)$ is the usual maximum modulus function of f. Show that

$$f(z) = \sigma(z) \sum_{(m,n) \in \mathbb{Z}^2} f(m+in) \frac{1}{\sigma'(m+in)(z-m-in)}$$

$$= \sigma(z) \sum_{(m,n) \in \mathbb{Z}^2} f(m+in) \frac{(-1)^{m+n+mn} e^{-\pi(m^2+n^2)/2}}{z-m-in}$$

uniformly on compact subsets of $\mathbb{C}$. Hint: use the contour integral method with the square contour C_p, $p \in \mathbb{N}$ described in §9.2. (J.M. Whittaker 1935, p. 72).

9.4 Let f satisfy the conditions of Problem 9.3. Show that

$$
f(z) = \sigma^2(z/\sqrt{2}) \sum_{(m,n)\in\mathbb{Z}^2} \left\{ \frac{f(\sqrt{2}(m+in))}{(z - \sqrt{2}(m+in))^2} \right.
$$
$$
\left. - 2\pi(m-in)\frac{f(\sqrt{2}(m+in))}{z - \sqrt{2}(m+in)} + \frac{f'(\sqrt{2}(m+in))}{z - \sqrt{2}(m+in)} \right\} e^{-\pi(m^2+n^2)}.
$$

(Higgins 1991).

10

IRREGULAR SAMPLING

A sampling process employing sample points that are not equidistantly spaced is usually called *irregular*, or sometimes *non-uniform*. Irregular sampling arises mathematically by simply asking the question "What is special about equidistantly spaced sample points?"; and then finding, as we shall in this chapter, that the answer is "Within certain limitations, nothing at all".

In practice it is often said that irregular sampling is the norm rather than the exception. For example, sample points are likely to be disturbed by "noise", meaning any random influences beyond our control. Again, it might be convenient to reconstruct a function from the locations of its zeros, or places where it crosses some datum function. Such points are, of course, intrinsic to the object function.

Again, sample points might be required to have a deterministic distribution, and then we ask if all functions of a given class can be reconstructed from their samples at these points. This is the only kind of irregular sampling we shall discuss here. A review of the applied aspects of irregular sampling can be found in Marvasti (1993), and there are many interesting studies on this theme in the Conference Proceedings of the 1995 workshop on sampling theory and applications, Jurmala, Latvia, 1995, published by the Institute of Electronics and Computer Science, Riga, Latvia.

A specialized framework for irregular sampling procedures involving the factorization of convolution operators has been developed in recent years by Feichtinger and Gröchenig; the reader is referred to Zayed (1993b, §9.1) for an expository account of this work. Different methods can be found in Zwaan (1990a).

10.1 Sets of stable sampling, of interpolation and of uniqueness

In any sampling process the mechanism for collecting each sample must consume a certain amount of the resources available to the process, including a certain period of time. Ideally we would like to ensure that the sample points are spaced far enough apart so that the mechanism for collecting one sample does not interfere with that of collecting another. On practical grounds it is usual, therefore, to make an assumption of the following kind.

Definition 10.1 *A countable set of real points $\{\lambda_n\}$ is called uniformly discrete, or sometimes separated, if its members are separated by at least some fixed positive distance l; so that*

$$0 < l \leq |\lambda_n - \lambda_m|, \qquad n \neq m.$$

From now on we shall assume that our sets of sample points $\{\lambda_n\}$ are always uniformly discrete, so that the density in Definition 10.3 always exists.

In the the following definition we consider the one-dimensional "multi-band" Paley–Wiener space PW_B introduced in Definition 6.19.

Definition 10.2 *We shall say that $\{\lambda_n\}$ is:*
1. A set of stable sampling for PW_B if $f \in PW_B \Rightarrow \|f\|_{L^2} \leq K\|f(\lambda_n)\|_{l^2}$ for some constant K which is independent of f.
2. A set of interpolation for PW_B if the moment problem $f(\lambda_n) = a_n$ for every n has a solution in PW_B whenever $(a_n) \in l^2$.
3. A set of uniqueness for PW_B if $f \in PW_B$ and $f(\lambda_n) = 0$ for every n implies that $f \equiv 0$.

The stability criterion ensures that errors in the output of a sampling and reconstruction process are bounded by errors in the input; again, sampling in practice is meaningless in the absence of such a condition. We refer simply to a *stable sampling series* as being one whose sample points $\{\lambda_n\}$ form a set of stable sampling.

Definition 10.3 *By $\sharp S$ we denote the cardinality of the set S. The density of a set of real points $\{\lambda_n\}$ is defined by*

$$D(\lambda_n) := \lim_{r \to \infty} \frac{\sharp\{\lambda_n : \lambda_n \in [-r, r]\}}{2r}.$$

The following theorem incorporates some very important results of Landau (1967b); the proof is postponed until Chapter 17.

Theorem 10.4 *Let $\{\lambda_n\}$ be a set of stable sampling. Then its density D satisfies*

$$D(\lambda_n) \geq m(B)/2\pi.$$

If $\{\lambda_n\}$ is a set of interpolation,

$$D(\lambda_n) \leq m(B)/2\pi.$$

Evidently a set of stable sampling is also a set of uniqueness. The converse is not generally true, however; the previous theorem shows that the densities of sets of stable sampling are bounded away from zero, but there exist sets of uniqueness with arbitrarily small density (Landau 1967a, p. 1702).

Definition 10.5 *We shall use the name Nyquist–Landau (N–L for short) for the critical density $m(B)/2\pi$; it is that sampling rate below which stable reconstruction is not possible.*

When B consists of a single interval $[-\pi w, \pi w]$ the N–L density coincides with that of Nyquist, namely w samples per second, or twice the highest frequency.

In the multi-band case it is important to note that the N–L density is less that the Nyquist density.

It is evident from Theorem 10.4 that only when $\{\lambda_n\}$ is distributed at the N–L density can it be both a set of stable sampling and a set of interpolation.

Theorem 10.6 *The set $\Lambda := \{\lambda_n\} \subset \mathbb{R}$, $n \in \mathbb{Z}$, is a set of uniqueness for $PW_{\pi w}$ if and only if $\{e^{ix\lambda_n}\}$ is complete in $L^2(-\pi w, \pi w)$.*

Proof We have

$$f(\lambda_n) = \frac{1}{\sqrt{2\pi}} \int_{-\pi w}^{\pi w} f^\wedge(x) e^{ix\lambda_n}\, dx, \qquad n \in \mathbb{Z}.$$

In one direction, if $\{e^{ix\lambda_n}\}$ is complete and $f(\lambda_n) = 0$, $n \in \mathbb{Z}$, then $\overline{f^\wedge}$, hence $f^\wedge$ and hence f, is null.

In the other direction, suppose g is an arbitrary member of $L^2(-\pi w, \pi w)$, and put

$$f(t) = \int_{-\pi w}^{\pi w} \overline{g(x)} e^{ixt}\, dx \in PW_{\pi w}.$$

Now $f(\lambda_n) = 0$ implies that $\langle g, e^{i\cdot\lambda_n}\rangle = 0$, and if this holds for every n and implies that f is null, then g must be null and the required completeness follows.

If $\{e^{ix\lambda_n}\}$ is merely assumed to be complete in $L^2(-\pi w, \pi w)$, the result of this theorem is all that we can expect, and it represents the weakest property that has any significance for sampling theory.

Indeed, if two members g and h of $PW_{\pi w}$ agree at the points $\{\lambda_n\}$ so that their difference is zero there, and if these points constitute a set of uniqueness, then h and g agree identically. This means that samples taken at a set of uniqueness determine members of $PW_{\pi w}$ uniquely, and while this is undoubtedly important, it does not lead to any process by which we can construct a function from its samples. Indeed, as Landau (1967a, p. 1702) points out, "The reason is that we know nothing about the effect on the reconstruction process of an error in measuring the sample values." Landau goes on to maintain that only stable sampling is meaningful in practice.

Even if we assume linear independence as well as completeness for $\{e^{ix\lambda_n}\}$, things are not much better. We have seen (in Problem 3.1) that such a set may fail to be a basis so that there is, in general, no series expansion in such a set. We could, however, pass to the time domain and apply the Gram–Schmidt orthogonalization process to obtain an expansion theory (as in Brown 1993, p. 67 using *finite linear independence*),[1] but because the coefficients would now contain linear combinations of samples, there is still no guarantee of stability.

Throughout Section 10.2 to follow, it will emerge that it is only when we can find conditions under which our set of exponentials $\{e^{ix\lambda_n}\}$ is a *basis* for $L^2(-\pi w, \pi w)$, or more desirably a *Riesz basis*, that matters take a more hopeful turn.

---

[1] That is, every finite subset of the set of complex exponentials is linearly independent.

10.2   Irregular sampling at minimal rate

Theorem 10.7 *Let $\{\lambda_n\} \subset \mathbb{R}$ be a uniformly discrete sequence of points, and suppose that the sampling expansion*

$$f(t) = \sum f(\lambda_n) S_n(t), \qquad t \in \mathbb{R},$$

holds for every $f \in PW_B$, and that (S_n) is a Riesz basis for PW_B. Then:
(i) $\{\lambda_n\}$ is both a set of stable sampling and a set of interpolation for PW_B.
(ii) $\{\lambda_n\}$ has the N–L density $m(B)/2\pi$.

Proof Let $(S_n^\star)$ be biorthonormal to (S_n); it is a Riesz basis, and we have $f(\lambda_n) = \langle f, S_n^\star \rangle$. Now the left half of Property 2 (Properties of Riesz bases, §3.2) shows that $\sum |f(\lambda_n)|^2$ is convergent, and the right half gives the stable sampling property.

That $\{\lambda_n\}$ is a set of interpolation for PW_B follows at once from Property 3 (Properties of Riesz bases, §3.2). This proves part (i). Part (ii) follows from Theorem 10.4.

When can we expect a sampling series like that in Theorem 10.7 to hold? To answer this, at least in the "low-pass" case, that is for the Paley–Wiener space $PW_{\pi w}$, we can first refer to Lemma 3.24 which guarantees that the biorthonormal set for $\{S_n\}$ is $\{k(\cdot, \lambda_n)\}$ where k is the reproducing kernel. In the present case this is $k(s,t) = w \operatorname{sinc} w(s - t)$ (from §6.7), and since $(\operatorname{sinc} w(\cdot - \lambda_n))^\wedge(x) = e^{-i\lambda_n x}/w\sqrt{2\pi}$, our question is, effectively: For what point sets $\{\lambda_n\}$ is $\{e^{i\lambda_n x}\}$ a Riesz basis for $L^2(-\pi w, \pi w)$?

A useful sufficient condition for this Riesz basis property to hold will be found in Theorem 10.9 below, but a characterization of such sets is a very much deeper problem. It was solved by Pavlov in 1979 (a comprehensive discussion of this problem, with references, can be found in Hruščev *et al.* 1981). To give something of its flavour, the characterization is in terms of an entire generating function

$$F(z) = \lim_{R \to \infty} \prod_{|\lambda_n| < R} \left(1 - \frac{z}{\lambda_n + i\delta} \right),$$

where δ is a real number chosen so that the points $\lambda_n + i\delta$ lie in a strip above the real axis. The characterization is stated here for real points λ_n so that any positive delta will work. Pavlov's necessary and sufficient condition for $\{e^{i\lambda_n x}\}$ to form a Riesz basis in $L^2(-\pi, \pi)$ is that $\{\lambda_n\}$ be uniformly discrete, F be of exponential type π, and that when x is real, $|F(x)|^2$ must satisfy Muckenhoupt's condition. A function f satisfies Muckenhoupt's condition if

$$\sup_{I \in \mathfrak{I}} \frac{1}{m(I)} \int_I f(x) \, dx \, \frac{1}{m(I)} \int_I \frac{1}{f(x)} \, dx < \infty,$$

where $\mathfrak{I}$ is the family of all real intervals I.

With this background we can now introduce an irregular sampling theorem. For convenience we shall denote the characteristic function of the interval $[-\pi w, \pi w]$ by $\chi_{\pi w}$.

Theorem 10.8 *Let* $\{\lambda_n\}$, $n \in \mathbb{Z}$, *be such that* $\{e^{i\lambda_n x}\}$ *is a Riesz basis for* $L^2(-\pi w, \pi w)$, *and denote the dual basis by* $\{\varphi_n^*(x)\}$. *Then:*
(i) $\{\lambda_n\}$ *is a set of stable sampling and of interpolation for* $PW_{\pi w}$.
(ii) *In norm and globally uniformly over* $\mathbb{R}$ *we have*

$$f(t) = \sum_{n \in \mathbb{Z}} f(\lambda_n) S_n(t), \qquad f \in PW_{\pi w}, \qquad (10.2.1)$$

where $S_n(t) = \sqrt{2\pi} \left(\overline{\varphi_n^*} \chi_{\pi w}\right)^{\vee}(t)$.
(iii) *The dual expansion is*

$$f(t) = w \sum_{n \in \mathbb{Z}} \langle f, S_n \rangle \operatorname{sinc} w(t - \lambda_n). \qquad (10.2.2)$$

Proof Part (i) follows from Theorem 10.7. As to parts (ii) and (iii), we find first that, since

$$\{e^{i\lambda_n x}, \varphi_n^*(x)\}$$

is a Riesz basis pair for $L^2(-\pi w, \pi w)$, then so is

$$\{e^{-i\lambda_n x}, \overline{\varphi_n^*}(x)\}.$$

The complex conjugate basis pair is preferred here so that we can connect with the reproducing equation for $PW_{\pi w}$, which is, from (6.7.5),

$$f(t) = w \int_{\mathbb{R}} f(s) \operatorname{sinc} w(s - t)\, ds.$$

By Fourier duality,

$$\{(e^{-i\lambda_n \cdot} \chi_{\pi w})^{\vee}(t), (\overline{\varphi_n^*} \chi_{\pi w})^{\vee}(t)\} \qquad (10.2.3)$$

is a Riesz basis pair for $PW_{\pi w}$.

First we expand $f \in PW_{\pi w}$ in the second of the bases (10.2.3). Then

$$f(t) = \sum_{n \in \mathbb{Z}} \langle f, (e^{-i\lambda_n \cdot} \chi_{\pi w})^{\vee} \rangle \left(\overline{\varphi_n^*} \chi_{\pi w}\right)^{\vee}(t).$$

Now, by a direct calculation using Appendix A no. 3a and no. 8,

$$(e^{-i\lambda_n \cdot} \chi_{\pi w})^{\vee}(t) = \sqrt{2\pi} w \operatorname{sinc} w(t - \lambda_n),$$

so that

$$\langle f, (e^{-i\lambda_n \cdot} \chi_{\pi w})^{\vee} \rangle = \sqrt{2\pi} f(\lambda_n),$$

by comparison with the reproducing equation. This gives (10.2.1).

The dual expansion (10.2.2) is now found directly from the first of the bases in (10.2.3).

The Convergence Principle of §6.6 applies, and gives the uniformity of convergence.

From §3.6 it was inevitable that the reproducing kernel would be involved here. Indeed, samples of f are made to enter the series just because the coefficients are examples of the reproducing formula of $PW_{\pi w}$.

Both expansions (10.2.1) and (10.2.2) can be obtained from the biorthogonal form of Kramer's Lemma (Theorem 8.8), but the appeal to Fourier duality seems more direct.

It is interesting to compare the results of this theorem with the ordinary cardinal series expansion for $PW_{\pi w}$. In the ordinary case the sample points are uniformly spaced; here they are not, and the effect is to split the cardinal series into two different representations, one of which is a sampling series while the other is not but is still an expansion in sinc functions.

$$* \qquad * \qquad *$$

While Theorem 10.8 gives a theoretical answer to the question of irregular sampling expansions in Riesz bases for $PW_{\pi w}$, it is still desirable to have some means of calculating the reconstruction functions S_n. Theorem 10.11 below addresses this point, and shows that under certain hypotheses the series (10.2.1) is a Lagrange type series. It is based on Theorem 10.9, which gives a simple and intuitively appealing sufficient condition for a set of complex exponentials to be a Riesz basis (Kadec 1964); also on Theorem 10.10, a special case of a result of Levinson (1936, p. 927), which provides the associated biorthonormal set in terms of an infinite product.

Provided that certain further conditions are met we can make a final step towards the explicit calculation of S_n, by means of an evaluation of certain infinite products in terms of gamma functions. This will lead us to Example 10.12, and another example can be found in Problem 10.2.

Theorem 10.9 (Kadec's "One Quarter" Theorem) *Let $\{\lambda_n\} \subset \mathbb{R}$, $n \in \mathbb{Z}$, and*

$$|n - \lambda_n| \leq D < \tfrac{1}{4}; \tag{10.2.4}$$

then $\{e^{i\lambda_n x}\}$ is a Riesz basis for $L^2(-\pi, \pi)$, and $1/4$ is the best possible constant.

Theorem 10.10 *Let $\{\lambda_n\}$ satisfy the Kadec "one quarter" condition (10.2.4), and put*

$$L(t) = (t - \lambda_0) \prod_{n=1}^{\infty} \left(1 - \frac{t}{\lambda_n}\right)\left(1 - \frac{t}{\lambda_{-n}}\right).$$

Then S_n of Theorem 10.8 is given by

$$S_n(t) = \frac{L(t)}{L'(\lambda_n)(t - \lambda_n)}.$$

These theorems give us the following special case of Theorem 10.8.[2]

Theorem 10.11 *Let $\{\lambda_n\}$ satisfy the conditions of Theorem 10.9. Then every $f \in PW_1$ is represented in a series of the Lagrange form*

$$f(t) = \sum_{n \in \mathbb{Z}} f(\lambda_n) \frac{L(t)}{L'(\lambda_n)(t - \lambda_n)},$$

in norm and globally uniformly over $\mathbb{R}$.

It is sometimes possible to evaluate $L(t)$ in terms of gamma functions (as in Whittaker and Watson 1962, p. 238). For example, if λ_n is a rational function of n for every n, and there is a factorization

$$\left(1 - \frac{t}{\lambda_n}\right)\left(1 - \frac{t}{\lambda_{-n}}\right) = \frac{(n - a_1)\dots(n - a_k)}{(n - b_1)\dots(n - b_k)}, \tag{10.2.5}$$

then

$$L(t) = (t - \lambda_0) \prod_{m=1}^{k} \frac{\Gamma(1 - b_m)}{\Gamma(1 - a_m)}. \tag{10.2.6}$$

Example 10.12 Let $\lambda_0 = 0$, $\lambda_n = n + c^2/n$, $\lambda_{-n} = -\lambda_n$, $n \in \mathbb{Z}$, where $0 \le c < \frac{1}{2}$. Then (10.2.6) gives, after some reduction,

$$L(t) = t\left(\cos \pi t - \cos \pi \sqrt{t^2 - 4c^2}\right)/2\sinh^2 \pi c.$$

10.3 Frames and over-sampling

The substitution of "frame" for "Riesz basis" in Theorem 10.8 results in the very similar Theorem 10.13 below, but there is one fundamental difference; the

---

[2]The origins of this theorem can be traced back to seminal work of Paley and Wiener (1934, p. 115); see also Hardy (1941, p. 332).

density of the sample points $\{\lambda_n\}$, a set of real distinct points, is now greater than the N–L density (unless the frame $\{e^{i\lambda_n x}\}$ reduces to a Riesz basis), and because of Theorem 10.4 this means that $\{\lambda_n\}$ cannot be a set of interpolation. We do retain the stable sampling property, however.

Theorem 10.13 *Let* $\{\lambda_n\}$, $n \in \mathbb{Z}$, *be such that* $\{e^{i\lambda_n x}\}$ *is a frame for* $L^2(-\pi w, \pi w)$, *and denote the dual frame by* $\{\psi_n^\star(x)\}$. *Then:*
(i) $\{\lambda_n\}$ *is a set of stable sampling for* $PW_{\pi w}$.
(ii)   *In norm and globally uniformly over* $\mathbb{R}$ *we have*

$$f(t) = \sum_{n \in \mathbb{Z}} f(\lambda_n) R_n(t), \qquad f \in PW_{\pi w}, \tag{10.3.1}$$

where $R_n(t) = \sqrt{2\pi} \left(\overline{\psi_n^\star} \chi_{\pi w} \right)^{\vee}(t)$.
(iii) *The dual expansion is*

$$f(t) = w \sum_{n \in \mathbb{Z}} \langle f, R_n \rangle \operatorname{sinc} w(t - \lambda_n). \tag{10.3.2}$$

(iv)   *Among all possible sets of coefficients for expanding* f *in the frame* $\{R_n\}$, *the samples* $\{f(\lambda_n)\}$, *and only these, have the smallest* l_2 *norm.*

Proof Let $f \in PW_{\pi w}$, and put $f^\wedge = g$, where $\operatorname{supp} g = [-\pi w, \pi w]$. The frame condition is

$$\sum_{n \in \mathbb{Z}} |\langle g, e^{i\lambda_n \cdot} \rangle|^2 \asymp \|g\|^2$$

from Definition 3.13, and by Plancherel's theorem this is equivalent to

$$\sum_{n \in \mathbb{Z}} |\langle f, (e^{i\lambda_n \cdot} \chi_{\pi w})^{\vee} \rangle|^2 \asymp \|f\|^2,$$

in which the norm and inner product are now those of $L^2(\mathbb{R})$. This is now a frame condition in $PW_{\pi w}$. The proof continues in much the same way as that of Theorem 10.8, using the two canonical frame expansions of Theorem 3.18. Part (iv) follows from the frame property in Problem 3.5.

After Theorem 10.7 it was essential to have a characterization of those point sets $\{\lambda_n\}$ for which the complex exponentials $\{e^{i\lambda_n x}\}$ formed a Riesz basis for the L^2-space over an interval. After Theorem 10.13 the same characterization problem arises, but this time for frames rather than Riesz bases. Again, a solution to this problem is known. First we need a definition.

Definition 10.14 *Let* $\Lambda = \{\lambda_n\}$, $n \in \mathbb{Z}$, *be uniformly discrete, and suppose there exist two positive numbers* d *and* L *such that*

$$\left| \lambda_n - \frac{n}{d} \right| \leq L, \qquad n \in \mathbb{Z}.$$

The Λ *is said to have uniform density* d.

Before coming to the promised characterization, Theorem 10.16 due to Jaffard, we shall quote the elegant sufficient condition for $\{e^{i\lambda_n x}\}$ to form a frame, due to Duffin and Schaeffer; both theorems can be found in Jaffard (1991).

Theorem 10.15 *Suppose that Λ has uniform density d, then $\{e^{i\lambda_n x}\}$ is a frame for $L^2(-\pi d, \pi d)$.*

Theorem 10.16 *Given Λ, there exists a bounded interval I such that Λ generates a frame of complex exponentials for $L^2(I)$ if and only if Λ can be written as the disjoint union of finitely many uniformly discrete subsequences, at least one of which has a uniform density.*

*   *   *

Sampling at densities above the N–L density is called *over-sampling* and has practical applications even though there is redundancy in the samples. Over-sampling can be a very effective technique in reducing errors that arise in the process of reconstruction from samples. For example, in compact disc technology as much as 8-fold over-sampling is sometimes used (Beaty *et al.* 1994, p. 440).

Example 10.17 (Regular Over-sampling) It follows from §3.1 that the set $\{(2\pi w)^{-1/2}e^{inx/w}\}$ is an orthonormal basis for $L^2(-\pi w, \pi w)$ with associated Parseval relation

$$\|f\|^2 = \sum_{n \in \mathbb{Z}} |c_n|^2,$$

where $c_n = \langle f, (2\pi w)^{-1/2}e^{in\cdot/w}\rangle$. Now define $g := f\chi_{\pi v}$ where $v < w$. Parseval's relation still holds; indeed, it is

$$\|g\|^2 = \sum_{n \in \mathbb{Z}} |d_n|^2,$$

where $d_n = \langle g, (2\pi w)^{-1/2}e^{in\cdot/w}\rangle$, but now the norm and inner product are those of $L^2(-\pi v, \pi v)$. But this shows that $\{(2\pi w)^{-1/2}e^{inx/w}\}$ is a tight frame for $L^2(-\pi v, \pi v)$ with unit frame bounds. Hence, Theorem 10.13 applies and the expansion for $PW_{\pi v}$ is just the ordinary cardinal series with sample points $\{n/w\}$; the density of these points is of course greater than that of $\{n/v\}$, and so we obtain regular over-sampling.

Several important properties of sets of complex exponentials can now be summarized in the following table:

For $L^2(-\pi w, \pi w)$ the set $\{e^{ix\lambda_n}\}$, $(n \in \mathbb{Z})$, is	For $PW_{\pi w}$ the set $\Lambda = \{\lambda_n\} \subset \mathbb{R}$, is
1. an orthonormal basis;	a regularly distributed set of stable sampling and of interpolation; $d(\Lambda) = m(B)/2\pi$;
2. a Riesz basis, equivalently an exact frame;	a set of stable sampling and of interpolation; $d(\Lambda) = m(B)/2\pi$;
3. a frame (possibly inexact);	a set of stable sampling; $d(\Lambda) \geq m(B)/2\pi$;
4. complete;	a set of uniqueness.

In any row the item in the left hand-column implies the item in the right-hand column. Another rule, needing rather liberal interpretation in a few places, is that in either column any item implies the items below it in that column.

PROBLEMS

10.1 Prove that if $\{h_n\}$, $n \in \mathbb{Z}$, is a tight frame for a Hilbert space H with unit frame bounds, and if $\|h_n\| = 1$ for every n, then $\{h_n\}$ is an orthonormal basis for H. This result does not apply to the frame in Example 10.17 above. Why not?

10.2 Let $\lambda_0 = \alpha$, $\lambda_n = n + \alpha$ and $\lambda_{-n} = -n + \beta$, $n \in \mathbb{N}$, and $0 \leq |\alpha|, |\beta| < \frac{1}{4}$. Then, for $f \in PW_1$,

$$f(t) = \frac{\Gamma(t - \alpha + 1)}{\Gamma(t - \beta + 1)} \sum_{n=0}^{\infty} f(n + \alpha) \frac{\Gamma(n + 1 + \alpha - \beta)}{n!} \text{sinc}(t - n - \alpha)$$

$$+ \frac{\Gamma(\beta - t)}{\Gamma(\alpha - t)} \sum_{m=0}^{\infty} f(-m + \beta) \frac{\Gamma(m + \alpha - \beta)}{(m - 1)!} \text{sinc}(t + m - \beta).$$

A slightly more general form of this series is attributed by Zayed (1993b, p. 148) to Grosjean, but no proof is given. Clearly it reduces to the "shifted" form of the cardinal series if $\alpha = \beta$ and to the ordinary form when $\alpha = \beta = 0$.

10.3 Show that the function set

$$\{1, \ \cos nx, \ \sin(n - \tfrac{1}{2})x\}, \qquad n \in \mathbb{N},$$

forms an orthonormal basis for $L^2(-\pi, \pi)$, and deduce a sampling theorem from it.

10.4 Suppose that $|\alpha| < 1/w$, and let $\{Tn \pm \alpha\}$, $n \in \mathbb{Z}$, $T = 2/w$, be a set of sample points. Show that for every $f \in PW_{\pi w}$,

$$f(t) = \frac{\cos \pi w \alpha - \cos \pi w t}{\pi w \sin \pi w \alpha} \sum_{n \in \mathbb{Z}} \left\{ \frac{f(Tn + \alpha)}{t - Tn - \alpha} - \frac{f(Tn - \alpha)}{t - Tn + \alpha} \right\}.$$

What can be said about the convergence and stability of this reconstruction?

More generally, when $T = m/w$ for some fixed $m \in \mathbb{N}$, the points

$$\lambda_{nk} := Tn + \alpha_k, \qquad n \in \mathbb{Z}, \ k = 1, \ldots, m, \ |\alpha_k| < T/2,$$

are called *bunched* sample points. Show that there is a sampling series for $f \in PW_{\pi w}$ using these points. This is a special case of a general sampling method due to Papoulis (1977, p. 194).

11

ERRORS AND ALIASING

The act of expanding a function into a sampling series is fraught with the possibility of incurring errors of various kinds. These are to be seen as deviations from the desired goal and usually arise from an incomplete knowledge of all the parameters; this is often due to our incapacity to measure with complete accuracy. Sometimes errors are incurred deliberately, especially if there is some trade-off value in doing so. Let us recall that we are always dealing with the case in which functions require infinitely many samples for exact reconstruction.

11.1 Errors

Several different kinds of error have been identified, and they have been studied intensively over the past few decades. The main ones are as follows.

The *truncation error* arises if one ignores all the samples outside a finite interval; it always occurs in practice of course. The object function cannot be reconstructed exactly because there is insufficient information. A more general kind of error occurs if samples are missing from *some* locations; it is called the *information loss error*.

If an available sample value differs from some assumed reference value (the "correct" value), then the difference between f produced using the available values and f produced using the "correct" values is called the *amplitude error*. In practice function values are usually calculated as an average over a small interval, and used instead of the unavailable "instantaneous values"; the error incurred is then called the *quantization error*.

A similar error arises from possible differences between available sample points and some "correct" set, and the resulting error is called the *time jitter error*. More information about jitter can be found in §11.2.

The *aliasing error* arises if one applies a sampling theorem to a function whose band-region is incompatible with that assumed in the theorem. Thus an aliasing error can arise from "under-sampling".[1]

All of these errors can occur in combination, of course.

The subject of errors in sampling theorems is a large, diverse field involving both stochastic and deterministic methods, and tends to belong to the more

---

[1] The aliasing phenomenon is graphically illustrated by a well-known anecdote. In old Western movies a stage coach is often seen moving at speed, but the wheels seem to be stationary or even revolving backwards. This distortion is an error due to aliasing, because the film's number of frames per second, the sampling rate, is not fast enough to account for the high frequencies generated by the rotating spokes.

specialized parts of the theory. The aliasing error is the only type that we shall cover here with any attempt at comprehensiveness. It has a character that is rather different from the others, and is the most intimately connected with the derivation of sampling theorems. An introduction to the subject will be found in §11.3 below, and furthermore it will be central to the treatment in Chapters 12 and 14.

Different methods are used to analyse other types of errors, and to gain something of their flavour we turn to the time jitter error in §11.2, and present just one theorem; it is the main deterministic theorem of the subject and is due to Butzer (1983). Its ingenious and rather intricate estimation procedures are typical of the proofs encountered in this part of error theory.

11.2 The time jitter error

As indicated in the last section, the time jitter error arises when we try to sample and reconstruct a function from samples taken at points which are perturbed from some spacing that is deemed *a priori* to be correct. Let $\{\delta_n\}$ denote a set of perturbation values.

Definition 11.1 *The time jitter error J_δ is given by*

$$(J_\delta f)(t) := f(t) - \sum_{n \in \mathbb{Z}} \left(\frac{n}{w} + \delta_n \right) \operatorname{sinc}(wt - n).$$

We always assume that the perturbation values $\{\delta_n\}$ are bounded; here it will be convenient to take

$$|\delta_n| \leq \delta \leq \min\{w^{-1}, e^{-1/2}\}, \qquad n \in \mathbb{Z}.$$

Without loss of generality we shall take $w \geq 1$ throughout this section.

We begin with two lemmas, both due to Splettstößer *et al.* (1981).

Lemma 11.2 *For $q > 1$ and as usual $1/p + 1/q = 1$,*

$$\sum_{n \in \mathbb{Z}} |\operatorname{sinc}(y - n)|^q < p, \qquad y \in \mathbb{R}.$$

Proof This sum is periodic with period 1, so that we need only consider $-\frac{1}{2} < y \leq \frac{1}{2}$. For such y, and for $n \neq 0$, we have

$$|\operatorname{sinc}(y - n)| = \left| \frac{\sin \pi(y - n)}{\pi(y - n)} \right| \leq \frac{1}{\pi |y - n|},$$

so that we can estimate our sum as follows.

$$\sum_{n \in \mathbb{Z}} |\operatorname{sinc}(y - n)|^q \leq 1 + \frac{1}{\pi^q} \sum_{n=1}^{\infty} \left\{ \frac{1}{(n - y)^q} + \frac{1}{(n + y)^q} \right\}$$

$$\leq 1 + \frac{1}{\pi^q} \sum_{n=1}^{\infty} \left\{ \frac{1}{(n-\frac{1}{2})^q} + \frac{1}{(n+\frac{1}{2})^q} \right\}$$

$$= 1 + \frac{1}{\pi^q} \left\{ \sum_{n=1}^{\infty} \frac{1}{(n+\frac{1}{2})^q} + \frac{1}{(1/2)^q} - \sum_{n=2}^{\infty} \frac{1}{(n-\frac{1}{2})^q} + 2 \sum_{n=2}^{\infty} \frac{1}{(n-\frac{1}{2})^q} \right\}$$

$$= 1 + \frac{1}{\pi^q} \left\{ 2^q + 2 \sum_{n=2}^{\infty} \frac{1}{(n-\frac{1}{2})^q} \right\}. \tag{11.2.1}$$

This last sum can be estimated by introducing an integral; thus

$$\sum_{n=2}^{\infty} \frac{1}{(n-\frac{1}{2})^q} < \int_1^{\infty} \frac{1}{(x-\frac{1}{2})^q}\, dx = \frac{2^{q-1}}{q-1}.$$

This means that (11.2.1) is bounded by

$$1 + \frac{1}{\pi^q} \left\{ 2^q + 2\frac{2^{q-1}}{q-1} \right\} = 1 + \left(\frac{2}{\pi} \right)^q \frac{q}{q-1}.$$

But since it can be shown (by elementary calculus methods, for example) that $(2/\pi)^q q \leq 1$, we can write

$$1 + \left(\frac{2}{\pi} \right)^q \frac{q}{q-1} \leq 1 + \frac{1}{q-1} = p,$$

and the proof is complete.

Lemma 11.3 *Let $f \in C(\mathbb{R})$ and suppose there exists a constant γ such that $0 < \gamma \leq 1$, and a constant M (depending on f) such that*

$$|f(t)| \leq M|t|^{-\gamma}, \qquad t \in \mathbb{R},\ t \neq 0. \tag{11.2.2}$$

Suppose further that R is a positive integer. Then whenever $p \geq 2/\gamma$ we have

$$\left\{ \sum_{|n|>R} \left| f\left(\frac{n}{w} \right) \right|^p \right\}^{1/p} < 2^{1/p} M w^{\gamma} R^{(1-p\gamma)/p}.$$

Proof Since $\gamma p \geq 2$ and $p \geq 2$, we have, directly from the hypotheses,

$$\left\{ \sum_{|n|>R} \left| f\left(\frac{n}{w} \right) \right|^p \right\}^{1/p} \leq M w^{\gamma} \left\{ 2 \sum_{n=R+1}^{\infty} n^{-\gamma p} \right\}^{1/p}$$

$$< M w^{\gamma} \left\{ \frac{2}{\gamma p - 1} \right\}^{1/p} R^{(1-\gamma p)/p} \leq 2^{1/p} M w^{\gamma} R^{(1-\gamma p)/p},$$

where once again the sum was estimated by an integral.

The error bound in the following theorem depends on the fact that f has a given rate of decay at infinity.

Theorem 11.4 *Let $f \in B_{\pi w}^1$, and let (11.2.2) hold. Then there exists a constant K, depending only on f, f' and γ, such that*

$$\|J_\delta f\|_C \le K\delta \log(1/\delta). \tag{11.2.3}$$

One can take

$$K = \frac{4}{\gamma}\left\{5^{\gamma/2}e\|f'\|_C + 2\cdot 2^{\gamma/2}Me^{1/4}\right\}. \tag{11.2.4}$$

Proof By Theorem 6.13 f has a pointwise representation in cardinal series, which we can substitute into the statement of Definition 11.1. Hölder's inequality can then be applied to obtain

$$|(J_\delta f)(t)| = \left|\sum_{n\in\mathbb{Z}}\left\{f\left(\frac{n}{w}\right) - f\left(\frac{n}{w}+\delta_n\right)\right\}\operatorname{sinc}(wt-n)\right|$$

$$\le \left\{\sum_{n\in\mathbb{Z}}\left|f\left(\frac{n}{w}\right) - f\left(\frac{n}{w}+\delta_n\right)\right|^p\right\}^{1/p}\left\{\sum_{n\in\mathbb{Z}}|\operatorname{sinc}(wt-n)|^q\right\}^{1/q}. \tag{11.2.5}$$

First, for any positive integer N we can decompose the first sum in (11.2.5) and estimate it using an appropriately arranged form of Minkowski's inequality (Theorem 2.4), as follows:

$$\left\{\left(\sum_{|n|<N}+\sum_{|n|\ge N}\right)\left|f\left(\frac{n}{w}\right)-f\left(\frac{n}{w}+\delta_n\right)\right|^p\right\}^{1/p}$$

$$\le\left\{\sum_{|n|<N}\left|f\left(\frac{n}{w}\right)-f\left(\frac{n}{w}+\delta_n\right)\right|^p\right\}^{1/p}$$

$$+\left\{\sum_{|n|\ge N}\left|f\left(\frac{n}{w}\right)-f\left(\frac{n}{w}+\delta_n\right)\right|^p\right\}^{1/p}. \tag{11.2.6}$$

It is now convenient to choose $N = b+1$ with

$$b = [\delta^{-1/\gamma}w^{\gamma p/(\gamma p-1)}] \ge 1, \tag{11.2.7}$$

where, as usual, [ ] means the "integer part", and the estimate follows from $\delta \le 1/w$, $w \ge 1$ and $\gamma p \ge 2$.

Now the first sum in (11.2.6) is bounded by

$$(2b+3)^{1/p} \sup_{n \in \mathbb{Z}} \sup_{t \in \mathbb{R}} |f(t) - f(t+\delta_n)|$$

and this in turn is bounded by

$$(2b+3)^{1/p}\delta \|f'\|_C,$$

where we have used the usual notation for the supremum norm, and the fact that f' exists everywhere on $\mathbb{R}$ by hypothesis.

To continue the estimations we note that $w^\gamma < 1/\delta$, since $0 < \gamma \le 1$, and so b can be estimated from (11.2.7). Hence

$$(2b+3)^{1/p} \le (5b)^{1/p} \le 5^{1/p}\delta^{-1/\gamma p}\delta^{-1/(\gamma p - 1)}$$

$$= 5^{1/p}\left(\frac{1}{\delta}\right)^{(2\gamma p - 1/)/\gamma p(\gamma p - 1)} \le 5^{1/p}\left(\frac{1}{\delta}\right)^{4/\gamma p}. \tag{11.2.8}$$

The factor 4 has been used here (when 3 would have been more precise) because a convenient value for p is needed, and this must be compatible with the requirement of Lemma 11.3. In fact, we can choose

$$p = \frac{4}{\gamma} \log \frac{1}{\delta},$$

a choice that is indeed not less than $2/\gamma$, and gives $(1/\delta)^{4/\gamma p} = e$. We now find from (11.2.8) that $(2b+3)^{1/p} \le 5^{\gamma/2}e$.

We have proved that the first sum in (11.2.6) is bounded by

$$5^{\gamma/2}e\delta \|f'\|_C.$$

As to the second sum in (11.2.6), we find from Lemma 11.3 that it is bounded by

$$2 \cdot 2^{1/p} M w^\gamma b^{(1-\gamma p)/\gamma p} \le 2 \cdot 2^{1/p} M w^\gamma \left(\delta^{-1/\gamma}w^{\gamma p/(\gamma p - 1)}\right)^{(1-\gamma p)/\gamma p}$$

$$\le 2 \cdot 2^{1/p} M \left(\frac{1}{\delta}\right)^{(1-\gamma p)/\gamma p} = 2 \cdot 2^{1/p} M\delta\left(\frac{1}{\delta}\right)^{1/\gamma p}.$$

Again, our choice of p (above) gives $(1/\delta)^{1/\gamma p} = e^{1/4}$, and we find that the second sum in (11.2.6) is bounded by

$$2 \cdot 2^{1/p} M\delta e^{1/4}.$$

Also, Lemma 11.2 shows that the second sum in (11.2.5) is bounded by p.

It only remains to collect these estimates together and we have reached the bound of (11.2.3); and the proof is complete.

11.3   The aliasing error

Let $\mathcal{A}_w$ denote the operation of expanding an arbitrary function f into its cardinal series, and let A denote the difference between f and the result of applying $\mathcal{A}_w$. Thus we have:

Definition 11.5 *Let*

$$(\mathcal{A}_w f)(t) := \sum_{n \in \mathbb{Z}} f\left(\frac{n}{w}\right) \operatorname{sinc}(wt - n), \qquad (11.3.1)$$

then the aliasing error A is defined to be

$$A(t) = |f(t) - (\mathcal{A}_w f)(t)|. \qquad (11.3.2)$$

Butzer *et al.* (1988, p. 16) consider that the first treatment of an aliasing problem occurs in the work of de la Vallée Poussin in 1908, and in Butzer and Stens (to appear) one can find an interesting historical account of this work and its influence.

In the context of aliasing it is customary to refer to the band-region of a particular function as the *actual band-region*, and to the band-region associated with (11.3.1), which in the present instance is the interval $[-\pi w, \pi w]$, as the *assumed band-region*. A class F of functions appropriate to the study of aliasing is found in the next definition. Its members may be band-limited to actual band-regions that differ from the assumed band-region, or they may not be band-limited at all.

Definition 11.6 $F := \{f \colon f \in L^2(\mathbb{R}) \cap C(\mathbb{R}),\ f^{\wedge} \in L^1(\mathbb{R})\}.$

For example, we have $PW_{\pi w} \subset F$, $w > 0$. Another important fact about F concerns the pointwise representation of its members as inverse Fourier transforms. First, we find that if $f \in F$, then a priori $f^{\wedge} \in L^1 \cap L^2$; thus $(f^{\wedge})^{\vee}$ can be calculated in the sense of either L^1 or L^2 and the results must agree almost everywhere (this can be seen by exchanging $\mathcal{F}$ and $\mathcal{F}^{-1}$ in Definition (2.12)). Hence, for $f \in F$,

$$f(t) = \frac{1}{\sqrt{2\pi}} \int_{\mathbb{R}} f^{\wedge}(x) e^{ixt}\, dx, \qquad (11.3.3)$$

and this is true for every $t \in \mathbb{R}$ since f is continuous.

A classical result giving a bound on the aliasing error A is the following theorem, stated originally by P. Weiss in 1963 and later proved by Brown (1967), who also supplied the extremal X.

Theorem 11.7 *Let $f \in F$ and let $\mathcal{A}_w$ be given by (11.3.1). Then*

$$A(t) \leq \frac{2}{\sqrt{2\pi}} \int_{|x| > \pi w} |f^{\wedge}(x)|\, dx. \qquad (11.3.4)$$

If f is real then this bound reduces to

$$A(t) \leq \sqrt{\frac{8}{\pi}} \int_{\pi w}^{\infty} |f^{\wedge}(x)| \, dx.$$

An extremal for (11.3.4) is

$$X(t) = \mathrm{sinc}(2wt - 1).$$

One says that X is *extremal* in the sense that there exists at least one value of t for which equality holds in (11.3.4) when X is expanded by (11.3.1).

The aliasing error bound in the previous theorem has been generalized in several directions in the research literature. We shall encounter one of these in Theorem 11.10 (below) and more in Chapters 12 and 14. The proof in the multi-dimensional case (§14.3) imitates the one-dimensional case very closely, so a separate proof will not be given here.

Demonstrating the sharpness of aliasing error bounds tends to be a rather *ad hoc* affair, involving some ingenuity in constructing extremals. Before discussing extremals in more detail, we come to a simple but significant corollary of the previous theorem; it gives an "approximate" representation in cardinal series for functions that are not necessarily band-limited.

Corollary 11.8 *Let $f \in F$. Then*

$$f(t) = \lim_{w \to \infty} \sum_{n \in \mathbb{Z}} f\left(\frac{n}{w}\right) \mathrm{sinc}(wt - n).$$

Proof The proof is imediate from (11.3.4), on recalling that $f^{\wedge} \in L^{1}(\mathbb{R})$.

$$* \qquad * \qquad *$$

The extremal X is our next point of discussion. This function belongs to $PW_{2\pi w}$ and hence to $PW_{(1+\alpha)\pi w}$ for every $\alpha \geq 1$, but not to $PW_{(1+\alpha)\pi w}$ for $\alpha < 1$. This raises the question as to whether there is an extremal in every space $PW_{(1+\alpha)\pi w}$, $0 < \alpha < 1$. The following corollary gives an affirmative answer.

Corollary 11.9 *Let $0 < \alpha < 1$. Then the inequality (11.3.4) is sharp for $f \in PW_{(1+\alpha)\pi w}$, equality holding when*

$$X_{\alpha}(t) := \mathrm{sinc}\left(\frac{\alpha}{2}(2wt - 1)\right) \sin \pi w t.$$

Proof First we note that X_α belongs to $PW_{(1+\alpha)\pi w}$. This follows from the Fourier transform of X_α, which is calculated using Appendix A, nos. 2, 3a and 8. The result is

$$X_\alpha^\wedge(x) = \begin{cases} -\dfrac{1}{\sqrt{2\pi}}\dfrac{1}{2iw\alpha}e^{-i(x+\pi w)/2w}, & -(1+\alpha)\pi w < x < -(1-\alpha)\pi w; \\[3mm] \dfrac{1}{\sqrt{2\pi}}\dfrac{1}{2iw\alpha}e^{-i(x-\pi w)/2w}, & (1-\alpha)\pi w < x < (1+\alpha)\pi w. \end{cases}$$

From (11.3.4),

$$A(t) \leq \frac{2}{\sqrt{2\pi}}\int_{\pi w < |x| < (1+\alpha)\pi w} |X_\alpha^\wedge(x)|\,dx$$

$$= \frac{2}{\sqrt{2\pi}}\frac{1}{\sqrt{2\pi}}\frac{1}{2w\alpha}\left\{\int_{-(1+\alpha)\pi w}^{-\pi w} dx + \int_{\pi w}^{(1+\alpha)\pi w} dx\right\}$$

$$= \frac{1}{2\pi w\alpha}(\pi w\alpha + \pi w\alpha) = 1.$$

On the other hand, we evidently have $X_\alpha(n/w) = 0$, $n \in \mathbb{Z}$, and $\max_{t\in\mathbb{R}} X_\alpha(t) = 1$, attained when $t = 1/2w$. So the requirement for an extremal is satisfied.

The details of this corollary are typical of the calculations required to demonstrate sharpness of aliasing error bounds. For example, it is convenient to search for extremals among functions that vanish at every sample point so that they are annihilated by $\mathcal{A}_w$; this is a great help when calculating A.

An interesting variation on the foregoing theme is found in the next theorem, which is due to Standish (1967). In this theorem we depart from our usual assumption that convergence should hold in the sense of symmetric partial sums. Convergence in (11.3.6) must be understood in the "two-way" sense; this is made explicit in the proof.

Theorem 11.10 *Let Φ be of bounded variation on $\mathbb{R}$, continuous at the points $(2k+1)\pi w$, $k \in \mathbb{Z}$, and normalized to be right continuous everywhere. Let f be the inverse Fourier–Stieltjes transform of Φ; that is,*

$$f(t) = \frac{1}{\sqrt{2\pi}}\int_\mathbb{R} e^{ixt}\,d\Phi(x). \tag{11.3.5}$$

Then the aliasing operator $\mathcal{A}_w$ given by

$$(\mathcal{A}_w f)(t) := \sum_{n\in\mathbb{Z}} f\left(\frac{n}{w}\right)\text{sinc}(wt - n) \tag{11.3.6}$$

converges for every $t \in \mathbb{R}$, and

$$A(t) = |f(t) - (\mathcal{A}_w f)(t)| \leq \frac{2}{\sqrt{2\pi}}\int_{|x| > \pi w} |d\Phi(x)|. \tag{11.3.7}$$

Proof Let

$$S_{M,N} := \sum_{n=-M}^{N} f\left(\frac{n}{w}\right) \mathrm{sinc}(wt - n), \qquad M, N \in \mathbb{N},$$

and in this partial sum let us replace $f(n/w)$ from (11.3.5) to obtain

$$S_{M,N} = \frac{1}{\sqrt{2\pi}} \int_{\mathbb{R}} \sum_{n=-M}^{N} e^{ixn/w} \mathrm{sinc}(wt - n) \, d\Phi(x).$$

The integrand is recognized as a partial sum of the Fourier series of $\left(e^{ixt}\right)_p$, where the subscript p denotes periodic extension (in x) with period $2\pi w$.

Now e^{ixt} is of bounded variation in $[-\pi w, \pi w)$ so that the partial sums of its Fourier series are bounded uniformly in M and N for each $t \in \mathbb{R}$. Furthermore, Φ can be written as the difference $\Phi_1 - \Phi_2$ of two non-decreasing functions, each with the continuity properties assumed for Φ. Hence each partial sum $S_{M,N}$ is bounded almost everywhere in x with respect to the measures induced by Φ_1 and Φ_2. The Lebesgue dominated convergence theorem applies, and we obtain

$$(\mathcal{A}_w f)(t) = \sum_{n \in \mathbb{Z}} f\left(\frac{n}{w}\right) \mathrm{sinc}(wt - n)$$

$$= \frac{1}{\sqrt{2\pi}} \int_{\mathbb{R}} \left(\lim_{M,N \to \infty} S_{M,N}\right) d\Phi(x) = \frac{1}{\sqrt{2\pi}} \int_{\mathbb{R}} \left(e^{ixt}\right)_p d\Phi(x).$$

This establishes the convergence. Finally,

$$|A(t)| = \frac{1}{\sqrt{2\pi}} \left| \int_{\mathbb{R}} \left\{ e^{ixt} - \left(e^{ixt}\right)_p \right\} d\Phi(x) \right|$$

$$\leq \frac{1}{\sqrt{2\pi}} \int_{|x| > \pi w} \left| e^{ixt} - \left(e^{ixt}\right)_p \right| |d\Phi(x)|$$

$$\leq \frac{2}{\sqrt{2\pi}} \int_{|x| > \pi w} |d\Phi(x)|.$$

Example 11.11 Let H_a denote the unit step function

$$H_a(x) := \begin{cases} 0, & x < a; \\ 1, & x \geq a, \end{cases}$$

where a is a real number. Let us take $w = 1$ in Theorem 11.10 and write $\mathcal{A}$ for $\mathcal{A}_1$, and let us take Φ to be H_a with $|a| < \pi$. Now $H_a' = \delta(x - a)$, so that

$$f(t) = \int_{\mathbb{R}} e^{ixt} \, dH_a(x) = \int_{\mathbb{R}} e^{ixt} \delta(x - a) \, dx = e^{iat},$$

and $f(n) = e^{ian}$. Hence

$$(\mathcal{A}f)(t) = \sum_{n \in \mathbb{Z}} e^{ian} \operatorname{sinc}(t - n). \tag{11.3.8}$$

Now, from the proof of Theorem 11.10, we have

$$|A(t)| \leq \frac{2}{\sqrt{2\pi}} \int_{|x|>\pi} \delta(x - a)\,|dx| = 0,$$

since $|a| < \pi$. This means that the series (11.3.8) converges to e^{iat}, and we have come to the result of Example 6.23 via a different route.

There is more to this example, however. Suppose we now make the same construction, but this time with $a = \pi$. We have

$$(\mathcal{A}f)(t) = \sum_{n \in \mathbb{Z}} (-1)^n \operatorname{sinc}(t - n) = \frac{\sin \pi t}{\pi} \sum_{n \in \mathbb{Z}} \frac{1}{t - n},$$

and this series is divergent (in the sense of Theorem 11.10). The choice $a = \pi$ was made in order to violate the continuity requirement on Φ, hence the theorem can be false without this requirement.

The reader wishing to read more about error analysis in this context might consult Jerri (1977, 1993), Butzer *et al.* (1988, §§3.5 and 5.4), Zwaan (1990a, 1995), Zayed (1993 §3.8) and their bibliographies.

PROBLEMS

11.1 Show that the (non-band-limited) function $e^{-|t|}$ does not differ from the sum of its cardinal series by more than $\frac{4}{\pi}(\frac{\pi}{2} - \arctan \pi)$ (take $\Phi(x)$ of Theorem 11.10 to be $1/(1 + x^2)$, and use Appendix A no. 23).

11.2 Show that an arbitrarily small violation of the bandwidth constraint supp $f^\wedge = [-\pi w, \pi w]$ is sufficient to make

$$\sum_{n \in \mathbb{Z}} f\left(\frac{n}{w}\right) \operatorname{sinc}(wt - n)$$

fail to converge, either pointwise or in the norm of $L^2(\mathbb{R})$ (Standish 1967).

11.3 Show that the aliasing operation $\mathcal{A}_w$ of §11.3 has some of the characteristics of a projection, in the following sense. First, define a more restricted domain for $\mathcal{A}_w$ by introducing

$$F_v := \{f : f \in F, \text{ supp } f^\wedge = [-\pi v, \pi v], \ f^\wedge \in BV[-\pi v, \pi v]\}, \qquad v > w,$$

where BV denotes the class of functions of bounded variation over $[-\pi v, \pi v]$. Then show that A_w is a bounded, linear and idempotent (that is, $A_w^2 = A_w$) transformation on F_v to $PW_{\pi w}$. (Beaty and Higgins 1994).

11.4 Show that a bound for the aliasing error $A(t)$ that is tighter than the one given in Theorem (11.7) is

$$\frac{2}{\sqrt{2\pi}} \int_{|x|>\pi w} |f^\wedge(x)| \, |\sin(\lambda(x)t/2)| \, dx,$$

where

$$\lambda(x) = \begin{cases} 0; & |x| \le \pi w, \\ 2\pi w \ \text{ceil}\left(\dfrac{|x| - \pi w}{2\pi w}\right) \text{sgn}\, x; & |x| > \pi w, \end{cases}$$

where ceil (ρ) denotes the smallest integer not less than the real number ρ. Hint: first find a formula for $(e^{ixt})_p$, where subscript p means periodic extension (in x) with period $2\pi w$, and calculate its Fourier series. (Stickler 1967).

12

SINGLE CHANNEL AND MULTI-CHANNEL SAMPLING

The reconstruction from samples of a band-limited function f is called *multi-channel* if the samples are not all taken from f itself, but at least some of them are taken from transformed versions of f. In practical terms, f has been passed through various processing channels before sampling. The name *single channel sampling* is used if only one processing channel is involved.

This idea goes back to the classic paper of Shannon (1949), in which the reconstruction of a band-limited signal from samples of the signal and of its derivatives is suggested. General methods for multi-channel sampling go back to work of Papoulis. This and later work in the area is synthesized in the expository account by Brown (1993). Concerning the need for multi-channel sampling, Brown says "In certain applications, data about a given band-limited signal may be available from several sources; for example we might have a spatial array of sensors such that each sensor receives a delayed version of the observed signal, the delay at a particular sensor being determined by the array geometry." This and ideas about derivative sampling suggest that "... a given lowpass signal may be processed by a parallel bank of m 'independent' linear, time invariant filters and then be completely reconstituted from the sampled outputs of these filters, each filter being sampled at $1/m$ times the input signal's Nyquist rate."

This raises an important point about sampling rates. In the multi-channel case the sample points can evidently occur at a density below the Nyquist density. For example if the sample points $\{2n/w\}$ in Theorem 12.8 (below) are compared to those in the ordinary single channel case, that is, $\{n/w\}$, it will be noticed that they occur at half the Nyquist density but there is twice the data to be collected at each point. The overall sampling rate, that is simply the rate at which samples must be observed, remains at the Nyquist value $1/w$. A similar remark would apply when there are more than two channels. Thus, in the case of multi-channel sampling it is often felt desirable to distinguish between the *sample point density* and the *sampling rate*. The reader is warned, however, that not all writers make this distinction. It remains an open question at the present time as to whether this multi-channel interpretation of the Nyquist rate is minimal in the same way as the ordinary Nyquist rate, that is, in the sense of Theorem 10.4.

In this chapter we shall present essentially two aspects of multi-channel sampling. One of these develops the idea of aliasing, introduced in the previous chapter; the other places two-channel, and by implication multi-channel, sampling in the context of expansions in Riesz bases, thereby gaining the advantages of this powerful and unifying method.

The means of establishing the Riesz basis property for certain sets of functions that lead to sampling series were developed by Rawn (1989) in the context of derivative sampling; the method was strengthened and extended by Higgins (1994), and will feature largely in this chapter and in the next.

12.1   Single channel sampling

Let $M(x)$ be a function defined on $[-\pi w, \pi w]$, of bounded variation there, and let $0 < 1/\beta \leq |M(x)| \leq \alpha < \infty$. Let $f \in PW_{\pi w}$, and let g be related to f by a similarity transformation $g = \mathcal{F}^{-1}\mathcal{M}\mathcal{F}f$, in which $\mathcal{F}$ denotes the Fourier transform as usual, and $\mathcal{M}$ denotes multiplication by $M(x)$.

In applications of signal theory g is said to be the *output*, when f is the *input*, to a *time invariant linear filter* with *transfer function* M. Thus

$$g(t) = \frac{1}{\sqrt{2\pi}} \int_{-\pi w}^{\pi w} M(x)f^{\wedge}(x)e^{ixt}\, dx; \qquad (12.1.1)$$

and we have

$$g\left(\frac{n}{w}\right) = \frac{1}{\sqrt{2\pi}} \int_{-\pi w}^{\pi w} f^{\wedge}(x)\varphi_n(x)\, dx, \qquad (12.1.2)$$

where we have put $\varphi_n(x) = M(x)e^{ixn/w}$.

We seek to represent f in terms of samples of g taken at the Nyquist sampling rate, in a series of the form

$$f(t) = \sum_{n \in \mathbb{Z}} g\left(\frac{n}{w}\right) S_n(t), \qquad (12.1.3)$$

where the "reconstruction functions" $\{S_n\}$ are to be determined.

A small technical point is of interest here. It is just because we use bases for $L^2(B)$ of the form $\{M(x)e^{ix\alpha n}\}$, $n \in \mathbb{Z}$, that samples are generated as coefficients in the series expansions that we shall be studying. This becomes evident in the proof of Theorem 12.1 below where samples of g appear, and the idea will recur in this chapter and the next, particularly in Theorems 12.4, 13.9 and 13.13.

Now the mapping $\mathcal{M}$ is a bounded linear invertible operator when restricted to $L^2(-\pi w, \pi w)$. This means that $\{\varphi_n\}$ is fully equivalent to the orthonormal basis $\{e^{ixn/w}/\sqrt{2\pi w}\}$ and is therefore, by Definition 3.7, a Riesz basis for $L^2(-\pi w, \pi w)$; and, furthermore, the dual basis $\{\varphi_n^*(x)\}$ can be written in the form $\{M^*(x)e^{ixn/w}\}$ with $M^* = (2\pi w\overline{M})^{-1}$, by the uniqueness of biorthonormal systems.

Theorem 12.1 *The sampling representation (12.1.3) holds for every $f \in PW_{\pi w}$ with $S_n(t) = \sqrt{2\pi}(\overline{M^*})^{\vee}(t - n/w)$.*

Proof Because $\{\varphi_n^*\}$ is a Riesz basis for $L^2(-\pi w, \pi w)$, it follows from Fourier duality that $\{(\overline{\varphi_n^*})^{\vee}\}$ is a Riesz basis for $PW_{\pi w}$. The expansion for f in this set takes the form

$$f(t) = \sum_{n \in \mathbb{Z}} \langle f, (\overline{\varphi_n})^{\vee} \rangle_{L^2(\mathbb{R})} (\overline{\varphi_n^*})^{\vee}(t) = \sum_{n \in \mathbb{Z}} \langle f, (\overline{\varphi_n})^{\vee} \rangle_{L^2(\mathbb{R})} (\overline{M^*})^{\vee} \left(t - \frac{n}{w} \right).$$

But by Plancherel's theorem and a simple calculation,

$$\langle f, (\overline{\varphi_n})^{\vee} \rangle_{L^2(\mathbb{R})} = \langle f^{\wedge}, \overline{\varphi_n} \rangle_{L^2(-\pi w, \pi w)} = \sqrt{2\pi} g \left(\frac{n}{w} \right),$$

from (12.1.2). Hence

$$f(t) = \sum_{n \in \mathbb{Z}} g \left(\frac{n}{w} \right) \sqrt{2\pi} \, (\overline{M^*})^{\vee} \left(t - \frac{n}{w} \right). \tag{12.1.4}$$

Of course, the Convergence Principle of §6.6 applies to (12.1.4).

The ordinary cardinal series is recovered from Theorem 12.1 by choosing $M \equiv 1$, whence $g \equiv f$.

The simplest non-trivial example of a one-channel series is where $\mathcal{M}$ denotes multiplication by $e^{-i\alpha x}$. Then $\mathcal{F}^{-1} \mathcal{M} \mathcal{F} = \tau_\alpha$, the shift operator $(\tau_\alpha f)(t) = f(t - \alpha)$. Now $\{ e^{-i\alpha x} e^{ixn/w} / \sqrt{2\pi w} \}$ is an orthonormal basis for $L^2(-\pi w, \pi w)$, and (12.1.4) becomes the "shifted" series

$$f(t) = \sum_{n \in \mathbb{Z}} f \left(\frac{n}{w} - \alpha \right) \operatorname{sinc} w \left(t - \frac{n}{w} + \alpha \right)$$

for Paley–Wiener functions f that we met in Example 8.3.

Example 12.2 Another simple example is obtained when we choose $M(x) \equiv -i \operatorname{sgn}(x)$, whence $g \equiv f^{\sim}$, the Hilbert transform of $f \in PW_{\pi w}$. One obtains

$$f(t) = - \sum_{n \in \mathbb{Z}} f^{\sim} \left(\frac{n}{w} \right) \operatorname{sinc} \frac{w}{2} \left(t - \frac{n}{w} \right) \sin \frac{\pi w}{2} \left(t - \frac{n}{w} \right). \tag{12.1.5}$$

A companion formula is easily obtained by using the operational formula $f^{\sim\sim} = -f$ (Appendix B), and the fact that if f belongs to $PW_{\pi w}$, then so does $f^{\sim}$. Then from (12.1.5) we obtain

$$f^{\sim}(t) = \sum_{n \in \mathbb{Z}} f \left(\frac{n}{w} \right) \operatorname{sinc} \frac{w}{2} \left(t - \frac{n}{w} \right) \sin \frac{\pi w}{2} \left(t - \frac{n}{w} \right).$$

We might note in passing that for $f \in PW_{\pi w}$ this formula could serve as the *definition* of the Hilbert transform!

A comprehensive discussion of these and other similar sampling schemes can be found in Stens (1983).

It will be recalled from Definition 11.6 that the class of functions $\{f : f \in L^2(\mathbb{R}) \cap C(\mathbb{R}), \ f^\wedge \in L^1(\mathbb{R})\}$ is denoted by F. Now let us denote by $\mathcal{A}_M f$ the operation of expanding f in the series (12.1.4) when we assume only that $f \in F$.

As in the previous chapter we have the notion of aliasing. We shall call $\mathcal{A}_M f$ the single channel *alias* of f; furthermore, let $A(t) = |f(t) - \mathcal{A}_M f(t)|$ denote the single channel *aliasing error*.

We shall now need a version of Poisson's summation formula suitable for functions of F.

Theorem 12.3 *Let $f \in F$. Then*

$$\sum_{n \in \mathbb{Z}} f\left(\frac{n}{w}\right) \frac{e^{-inv/w}}{\sqrt{2\pi w}} \sim \sqrt{w} \sum_{n \in \mathbb{Z}} f^\wedge(2n\pi w + v). \tag{12.1.6}$$

where, as in (2.3.1), " $\sim$ " indicates that the right-hand side of (12.1.6) converges in the norm of $L^1(-\pi w, \pi w)$ to a function whose formal Fourier series is the left-hand side of (12.1.6).

Proof The proof starts in a similar way to that of Theorem 2.21. We have

$$\int_{-\pi w}^{\pi w} \left| \sum_{n \in \mathbb{Z}} f^\wedge(2\pi wn + v) \right| dv \le \int_{-\pi w}^{\pi w} \sum_{n \in \mathbb{Z}} |f^\wedge(2\pi wn + v)| \, dv. \tag{12.1.7}$$

Now we can interchange the order of integration and summation on the right-hand side of (12.1.7) if the expression exists with either ordering. In fact, it is convenient to show existence in the other order:

$$\sum_{n \in \mathbb{Z}} \int_{-\pi w}^{\pi w} |f^\wedge(2\pi wn + v)| \, dv = \sum_{n \in \mathbb{Z}} \int_{(2n-1)\pi w}^{(2n+1)\pi w} |f^\wedge(x)| \, dx = \int_{\mathbb{R}} |f^\wedge(x)| \, dx < \infty.$$

This gives the required convergence, since $f^\wedge \in L^1(\mathbb{R})$. We now use the same idea on interchanging sum and integral to prove the second part; thus the kth Fourier coefficient of the right-hand side of (12.1.6) is

$$\sqrt{w} \int_{-\pi w}^{\pi w} \sum_{n \in \mathbb{Z}} f^\wedge(2\pi wn + v) e^{ivk/w} \, dv$$

$$= \sqrt{w} \sum_{n \in \mathbb{Z}} \int_{(2n-1)\pi w}^{(2n+1)\pi w} f^\wedge(x) e^{i(x - 2\pi wn)k/w} \, dx$$

$$= \sqrt{w} \sum_{n \in \mathbb{Z}} \int_{(2n-1)\pi w}^{(2n+1)\pi w} f^\wedge(x) e^{ixk/w} \, dx$$

$$= \sqrt{w} \int_{\mathbb{R}} f^\wedge(x) e^{ixk/w} \, dx = \sqrt{2\pi w} f\left(\frac{k}{w}\right).$$

The last equality is a consequence of (11.3.3), and this completes the proof.

For the next theorem we need to make a small modification to the multiplier operator $\mathcal{M}$. Because $f^\wedge$ is no longer required to be supported on an interval, we shall take $\mathcal{M}$ to denote multiplication by a factor $M(x)$ that is defined on all of $\mathbb{R}$. It is assumed to satisfy the same bounds as before, but this time for all real x.

Theorem 12.4 *Let $f \in F$. Then the single channel aliasing error satisfies*

$$|f(t) - \mathcal{A}_M f(t)| \le \frac{1 + \alpha\beta}{\sqrt{2\pi}} \int_{|x| > \pi w} |f^\wedge(x)|\, dx. \qquad (12.1.8)$$

Proof It is clear that $\mathcal{S}f := \mathcal{F}^{-1}\mathcal{M}\mathcal{F}f$ belongs to $L^2(\mathbb{R})$ if f belongs to $L^2(\mathbb{R})$. Furthermore, $(\mathcal{S}f)^\wedge = \mathcal{F}\mathcal{F}^{-1}\mathcal{M}\mathcal{F}f = M(x)f^\wedge(x) \in L^1(\mathbb{R})$, so that $\mathcal{S}f$ belongs to F if f belongs to F. Therefore we can apply (12.1.6) to $\mathcal{S}f$, and obtain

$$\sum_{n \in \mathbb{Z}} (\mathcal{S}f)\left(\frac{n}{w}\right) \frac{e^{-inv/w}}{\sqrt{2\pi w}} \sim \sqrt{w} \sum_{n \in \mathbb{Z}} (\mathcal{M}\mathcal{F}f)\,(2n\pi w + v). \qquad (12.1.9)$$

Because of Theorem 2.5 we can now legitimately multiply both sides of (12.1.9) by

$$\frac{(M(v))^{-1}e^{ivt}}{\sqrt{2\pi w}},$$

integrate over $[-\pi w, \pi w]$ (whose characteristic function is denoted $\chi_{\pi w}$) and obtain an equality. Then, on comparison with 12.1.4 and recalling that $g(n/w) = \mathcal{S}f(n/w)$, some calculations show that the left-hand side becomes

$$(\mathcal{A}_M f)(t) = \frac{1}{w\sqrt{2\pi}} \sum_{n \in \mathbb{Z}} (\mathcal{S}f)\left(\frac{n}{w}\right) (\chi_{\pi w} M^{-1})^\vee \left(t - \frac{n}{w}\right). \qquad (12.1.10)$$

The right hand-side becomes

$$\frac{1}{\sqrt{2\pi}} \sum_{n \in \mathbb{Z}} e^{-2\pi i n t w} \int_{(2n-1)\pi w}^{(2n+1)\pi w} (M(x - 2n\pi w))^{-1} M(x) e^{itx} f^\wedge(x)\, dx.$$

Now let us restructure

$$f(t) = \frac{1}{\sqrt{2\pi}} \int_{\mathbb{R}} e^{ixt} f^\wedge(x)\, dx$$

in the form

$$f(t) = \frac{1}{\sqrt{2\pi}} \sum_{n \in \mathbb{Z}} \int_{(2n-1)\pi w}^{(2n+1)\pi w} e^{ixt} f^\wedge(x)\, dx.$$

We now find from (12.1.10) that

$$|f - (\mathcal{A}_M f)(t)| =$$

$$\frac{1}{\sqrt{2\pi}} \left| \sum_{n \in \mathbb{Z}} \int_{(2n-1)\pi w}^{(2n+1)\pi w} e^{ixt} f^{\wedge}(x) \left\{ 1 - e^{-2\pi i n t w} (M(x - 2n\pi w))^{-1} M(x) \right\} dx \right|.$$

Clearly, the term corresponding to $n = 0$ vanishes, and the result follows on estimating the integrand in an obvious way and reconstituting the integral over $\mathbb{R}$.

The classic result, Theorem 11.7 is obtained if $f \in PW_{\pi w}$ and $M \equiv 1$. We also have:

Corollary 12.5 *Let $f \in PW_{\pi w}$ and $|M(x)| = 1$ a.e. Then the result of Corollary 7.9 is recovered, in that*

$$f(t) = (\mathcal{A}_M f)(t) = \frac{1}{w\sqrt{2\pi}} \sum_{n \in \mathbb{Z}} (\mathcal{S}f) \left(\frac{n}{w} \right) (M^{-1})^{\vee} \left(t - \frac{n}{w} \right) \qquad (12.1.11)$$

and the expansion functions form an orthonormal basis for $PW_{\pi w}$.

Proof Since $f \in PW_{\pi w}$ the aliasing error vanishes, so the expansion is valid. The orthonormal basis property follows from the fact that

$$(M^{-1})^{\vee} \left(t - \frac{n}{w} \right) = \left[(M(\cdot))^{-1} \chi_{\pi w}(\cdot) e^{i \cdot n/w} \right]^{\vee} (t),$$

that $\left\{ e^{ixn/w} \right\}_{n \in \mathbb{Z}}$ forms an orthonormal basis in $L^2(-\pi w, \pi w)$, and Plancherel's theorem.

Example 12.6 (Continuation of Example 12.2) Choose $M(x) = -i \operatorname{sgn} x$ again, and let $f \in F$. In Theorem 12.4 we have $\alpha = \beta = 1$, and so the aliasing error bound is the same as in the classic case, Theorem 11.7.

Another example of a multiplier M arises in the application of sampling theory in communications engineering. The simplest practical approach to signal reconstruction is to generate a step-function from the sample values and then smooth it by using a low-pass filter. This simple method of analogue-to-digital conversion goes by the name of "sample and hold". The process can be viewed, ideally, as a one-channel system with multiplier $M(x) = \operatorname{sinc}(x/2\pi w)$ (details can be found in Beaty *et al.* (1994)). It is interesting to note that for $f \in PW_{\pi w}$ the action of this multiplier is equivalent to taking a local average in the time domain over a window of length $1/4w$. The hypotheses of Theorem 12.1 are satisfied since

$$\frac{2}{\pi} \leq \frac{\sin(x/2w)}{x/2w} \leq 1,$$

when $x \in [-\pi w, \pi w]$. This means that a Paley–Wiener function can be recovered from samples of its local average; unfortunately, a closed form for the reconstruction function

$$\frac{1}{\sqrt{2\pi}} \int_{-\pi w}^{\pi w} \frac{x/2w}{\sin(x/2w)} e^{ixt} \, dx$$

does not seem to be known, although it could be found as a series involving central factorial numbers; for this purpose one could use an expansion for the reciprocal sinc function in Butzer and Stens (1992a, p. 168).

12.2   Two channels

Sampling through more than one channel requires certain modifications in the methods of the previous section. These will be described in the case of two channels; more than two channels can be treated similarly but at the price of some complexity in notation.

Let there be two functions g_1 and g_2 allied to the object function f, of the form

$$g_j(t) = \frac{1}{\sqrt{2\pi}} \int_{-\pi w}^{\pi w} M_j(x) f^\wedge(x) e^{ixt} \, dx, \qquad j = 1, 2.$$

Throughout this section we assume that the transfer functions M_j are defined on $[-\pi w, \pi w]$ and are bounded and measurable there; however, they do not both need to be bounded away from zero. The invertibility of the operator $\mathcal{M}$ of the previous section is replaced here with the requirement that the matrix M of Definition 12.7 below is non-singular.

In order to obtain a sampling theorem that will reconstruct a Paley–Wiener function from samples of g_1 and g_2 we need to identify an appropriate Riesz basis for $L^2(-\pi w, \pi w)$. For this purpose let

$$L^2(0, \pi w) \oplus L^2(0, \pi w) = \left\{ (\psi, \omega) : \psi, \omega \in L^2(0, \pi w) \right\}$$

be the direct sum (§3.5) with norm denoted by $\|(\psi, \omega)\|$, for which the set $\{w_n\}$, where

$$\left. \begin{aligned} w_{2m} &= \left(e^{ix2m/w}/\sqrt{\pi w}, \theta \right) \\ w_{2m+1} &= \left(\theta, e^{ix2m/w}/\sqrt{\pi w} \right) \end{aligned} \right\}, \qquad m \in \mathbb{Z}, \qquad (12.2.1)$$

is evidently an orthonormal basis.

Define the mapping

$$\mathcal{T} : L^2(0, \pi w) \oplus L^2(0, \pi w) \longmapsto L^2(-\pi w, \pi w)$$

by

$$\mathcal{T}(\psi, \omega) = M_1(x)\psi_p(x) + M_2(x)\omega_p(x),$$

where subscript p denotes extension by periodicity to all of $\mathbb{R}$; thus ψ_p and ω_p have period πw. In particular, let us put

$$\varphi_{2m} := \mathcal{T} w_{2m} = M_1(x)e^{ix2m/w}/\sqrt{\pi w},$$

$$\varphi_{2m+1} := \mathcal{T} w_{2m+1} = M_2(x)e^{ix2m/w}/\sqrt{\pi w}. \tag{12.2.2}$$

Here we can dispense with the subscript p since the exponentials already have the right period.

In the definition of the matrix M below, and throughout the rest of this section, we adopt the convention that if $G(x)$ is a function with domain $[-\pi w, \pi w]$, we can break up this domain into two halves and write $G = G^+ \chi_{[0,\pi w]} + G^- \chi_{[-\pi w,0]}$.

In extensions to the multi-channel case one would break up $[-\pi w, \pi w]$ into N equal sub-intervals (where N is the number of channels). This is a simple but very effective device in the proofs of multi-channel sampling theorems. With this convention in mind we now adopt:

Definition 12.7

$$M := \begin{pmatrix} M_1^+(x) & M_2^+(x) \\ M_1^-(x - \pi w) & M_2^-(x - \pi w) \end{pmatrix}, \qquad x \in [0, \pi w]. \tag{12.2.3}$$

When M is non-singular we denote the elements of M^{-1} (again for $x \in [0, \pi w]$) as follows:

$$M^{-1} = \pi w \begin{pmatrix} M_1^{*+}(x) & M_1^{*-}(x - \pi w) \\ M_2^{*+}(x) & M_2^{*-}(x - \pi w) \end{pmatrix}.$$

The matrix M now takes over the rôle of the single multiplier $M(x)$ of the single channel case.

Before coming to the next sampling theorem we need to identify $\mathcal{T}^*$, the adjoint of the transformation $\mathcal{T}$. Definition 3.8 suggests that we are looking for a pair (h_1, h_2) such that, given any $h \in L^2(-\pi w, \pi w)$, there is a transformation $\mathcal{T}^*$ such that

$$\mathcal{T}^* : h \mapsto (h_1, h_2) \in L^2(0, \pi w) \oplus L^2(0, \pi w)$$

and such that

$$\int_{-\pi w}^{\pi w} \left[\overline{M_1(x)} \overline{\psi_p}(x) + \overline{M_2(x)} \overline{\omega_p}(x) \right] h(x)\, dx$$

$$= \int_0^{\pi w} \overline{\psi}(x) h_1(x)\, dx + \int_0^{\pi w} \overline{\omega}(x) h_2(x)\, dx \tag{12.2.4}$$

where (ψ, ω) is any member of $L^2(0, \pi w) \oplus L^2(0, \pi w)$.

With the notational convention as before, put $h(x) = h^+ \chi_{[0,\pi w]} + h^- \chi_{[-\pi w,0]}$. After a simple change, the left-hand integral in (12.2.4) can be written

$$\int_0^{\pi w} \left\{ \overline{\psi}(x)[\overline{M_1^+}(x)h^+(x) + \overline{M_1^-}(x - \pi w)h^-(x - \pi w)] \right.$$

$$\left. + \overline{\omega}(x)[\overline{M_2^+}(x)h^+(x) + \overline{M_2^-}(x - \pi w)h^-(x - \pi w)] \right\} dx.$$

A comparison of this with the right-hand integral in (12.2.4) suggests that we take[1]

$$\begin{pmatrix} h_1(x) \\ h_2(x) \end{pmatrix} = \begin{pmatrix} \overline{M_1^+}(x) & \overline{M_1^-}(x - \pi w) \\ \overline{M_2^+}(x) & \overline{M_2^-}(x - \pi w) \end{pmatrix} \begin{pmatrix} h^+(x) \\ h^-(x - \pi w) \end{pmatrix},$$

and that we can define

$$(T^* h)(x) := (\overline{M})^T \begin{pmatrix} h^+(x) \\ h^-(x - \pi w) \end{pmatrix}, \qquad (12.2.5)$$

where superscript T denotes the transposed matrix.

Theorem 12.8 *Let M_1 and M_2 be bounded measurable functions defined on $[-\pi w, \pi w]$, and such that M is non-singular. Then the following facts ensue from the notations developed above:*

1. *The set $\{\varphi_n\}$, $n \in \mathbb{Z}$, of (12.2.2) is a Riesz basis for $L^2(-\pi w, \pi w)$.*
2. *The dual basis $\{\varphi_n^*\}$ is given by*

$$\left. \begin{array}{l} \{\varphi_{2m}^*\} = \sqrt{\pi w}\, \overline{M_1^*}(x) e^{ix2m/w} \\ \{\varphi_{2m+1}^*\} = \sqrt{\pi w}\, \overline{M_2^*}(x) e^{ix2m/w} \end{array} \right\}, \qquad m \in \mathbb{Z}.$$

3. *For every $f \in PW_{\pi w}$ we have*

$$f(t) = \sum_{n \in \mathbb{Z}} g_1\left(\frac{2n}{w}\right) R_n(t) + g_2\left(\frac{2n}{w}\right) S_n(t), \qquad (12.2.6)$$

where

$$R_n(t) = \sqrt{2\pi}\, (M_1^*)^\vee (t - 2n/w),$$
$$S_n(t) = \sqrt{2\pi}\, (M_2^*)^\vee (t - 2n/w).$$

The Convergence Principle of §6.6 applies.

Proof First, it is evident that T is a linear transformation. We now show that it is one-to-one and onto. Let $H \in L^2(-\pi w, \pi w)$ and $H := H^+ \chi_{[0, \pi w]} + H^- \chi_{(-\pi w, 0]}$. We want to find $\psi, \omega \in L^2(0, \pi w)$ such that

$$H(x) = M_1(x)\psi_p(x) + M_2(x)\omega_p(x).$$

It is therefore natural to take, for $x \in [0, \pi w]$,

$$\begin{pmatrix} \psi_p(x) \\ \omega_p(x) \end{pmatrix} = \begin{pmatrix} M_1^+(x) & M_2^+(x) \\ M_1^-(x - \pi w) & M_2^-(x - \pi w) \end{pmatrix}^{-1} \begin{pmatrix} H^+(x) \\ H^-(x - \pi w) \end{pmatrix}.$$

---

[1]Pairs will be written as column vectors when convenient.

Now since the M_j's are bounded then so are the elements of M^{-1}. Hence ψ_p and ω_p belong to $L^2(0, \pi w)$. Because M is non-singular it is clear that H and the pair (ψ, ω) determine each other uniquely, so that $\mathcal{T}$ is one-to-one and onto.

Now we must show that $\mathcal{T}$ is bounded. Let $m_i = \sup_{x \in [0, \pi w]} |M_i(x)|$, $i = 1, 2$. Then

$$
\begin{aligned}
\|\mathcal{T}(\psi, \omega)\|^2 &= \|M_1(x)\psi_p(x) + M_2(x)\omega_p(x)\|^2_{L^2(-\pi w, \pi w)} \\
&\le 4 \left\{ m_1 \|\psi\|_{L^2(0,\pi w)} + m_2 \|\omega\|_{L^2(0,\pi w)} \right\}^2 \\
&\le 8 \max (m_1^2, m_2^2) \left\{ \|\psi\|^2_{L^2(0,\pi w)} + \|\omega\|^2_{L^2(0,\pi w)} \right\} \\
&\le (\text{constant}) \|(\psi, \omega)\|^2.
\end{aligned}
$$

It follows that the set $\{\varphi_n\}$, $n \in \mathbb{Z}$ (in (12.2.2)) is fully equivalent to the orthonormal basis $\{w_n\}$ of (12.2.1). Therefore by property 4 of §3.2 it is a Riesz basis for $L^2(-\pi w, \pi w)$, and this completes the proof of part 1.

To prove part 2, we note first that the basis $\{\varphi_n^\star\}$ dual to $\{\varphi_n\}$ is associated with the adjoint transformation $\mathcal{T}^\star$ (as in Definition 3.8 and the discussion following it). Now

$$
\begin{aligned}
\varphi_{2m}^\star(x) = (\mathcal{T}^\star)^{-1} w_{2m}(x) &= (\overline{M}^{-1})^T w_{2m}(x) \\
&= \begin{pmatrix} \sqrt{\pi w}\, \overline{M_1^{\star+}}(x) e^{ix2m/w} \\ \sqrt{\pi w}\, \overline{M_1^{\star-}}(x - \pi w) e^{ix2m/w} \end{pmatrix}, \qquad x \in [-\pi w, \pi w],
\end{aligned}
$$

so that, because of the way the object function is decomposed in 12.2.5, the inversion gives finally

$$
\begin{aligned}
\varphi_{2m}^\star(x) &= \sqrt{\pi w}\, \overline{M_1^{\star+}}(x) e^{ix2m/w} \chi_{[0,\pi w]} + \sqrt{\pi w}\, \overline{M_1^{\star-}}(x - \pi w) e^{ix2m/w} \chi_{[-\pi w, 0]} \\
&= \sqrt{\pi w}\, \overline{M_1^\star}(x) e^{ix2m/w}, \qquad x \in [-\pi w, \pi w].
\end{aligned}
$$

A similar calculation holds for $\varphi_{2m+1}^\star$, and this proves part 2.

To prove part 3, we first expand $f^\wedge$ in the Riesz basis $\{\overline{\varphi_n^\star}\}$:

$$
f^\wedge(x) = \sum_{n \in \mathbb{Z}} \langle f^\wedge, \overline{\varphi_n} \rangle \overline{\varphi^\star}_n(x)
$$

$$
= \sum_{m \in \mathbb{Z}} \left\langle f^\wedge, \overline{\varphi_{2m}} \right\rangle \overline{\varphi^\star}_{2m}(x) + \sum_{m \in \mathbb{Z}} \langle f^\wedge, \overline{\varphi_{2m+1}} \rangle \overline{\varphi^\star}_{2m+1}(x) \qquad (12.2.7)
$$

using subseries convergence (property 5 of §3.2). Now

$$
\langle f^\wedge, \overline{\varphi_{2m}} \rangle = \int_{-\pi w}^{\pi w} f^\wedge(x) M_1(x) e^{ix2m/w} / \sqrt{\pi w}\, dx = \sqrt{2\pi} g_1 \left(\frac{2m}{w} \right) / \sqrt{\pi w},
$$

and similarly,

$$\langle f^\wedge, \overline{\varphi_{2m+1}} \rangle = \sqrt{2\pi} g_2 \left(\frac{2m}{w} \right) / \sqrt{\pi w}.$$

Also,

$$\left(\overline{\varphi_{2m}^\star} \right)^\vee (t) = \frac{\sqrt{\pi w}}{\sqrt{2\pi}} \int_{-\pi w}^{\pi w} M_1^\star(x) e^{-ix2m/w} e^{ixt} \, dx$$

$$= \frac{\sqrt{\pi w}}{\sqrt{2\pi}} \int_{-\pi w}^{\pi w} M_1^\star(x) e^{ix(t - 2m/w)} \, dx$$

$$= \sqrt{\pi w} \, (M_1^\star)^\vee (t - 2m/w),$$

and similarly for $\left(\overline{\varphi_{2m+1}^\star} \right)^\vee$. Then, the Fourier dual of (12.2.7) is (12.2.6).

Example 12.9 (Function and Hilbert Transform Sampling)   An example of the two-channel method is to show that the sampling series appropriate for reconstructing a Paley–Wiener function f from samples of both itself and its Hilbert transform $f^\sim$ is

$$f(t) = \sum_{n \in \mathbb{Z}} \left\{ f\left(\frac{2n}{w} \right) \cos \frac{\pi w}{2} \left(t - \frac{2n}{w} \right) \right.$$

$$\left. - f^\sim \left(\frac{2n}{w} \right) \sin \frac{\pi w}{2} \left(t - \frac{2n}{w} \right) \right\} \operatorname{sinc} \frac{w}{2} \left(t - \frac{2n}{w} \right), \quad (12.2.8)$$

for every $f \in PW_{\pi w}$.

Indeed, by inspecting Appendix A we find in item 5 the multiplier appropriate to the Hilbert transform. We are prompted to choose $M_1(x) \equiv 1$ and $M_2(x) \equiv -i \operatorname{sgn}(x)$ for our transfer functions. Then

$$M = \begin{pmatrix} 1 & -i \\ 1 & i \end{pmatrix}, \qquad M^{-1} = \frac{1}{2} \begin{pmatrix} 1 & 1 \\ i & -i \end{pmatrix},$$

$$M_1^\star(x) \equiv 1/2\pi w, \qquad M_2^\star(x) \equiv (i/2\pi w) \operatorname{sgn} x,$$

$$\begin{cases} \varphi_{2m} = e^{ix2m/w} / \sqrt{\pi w} \\ \varphi_{2m+1} = -i \operatorname{sgn} x \, e^{ix2m/w} / \sqrt{\pi w}, \end{cases}$$

and

$$\begin{cases} \varphi_{2m}^\star = \dfrac{\sqrt{\pi w}}{2\pi w} e^{ix2m/w} \\ \varphi_{2m+1}^\star = \dfrac{\sqrt{\pi w}}{2\pi w} (-i \operatorname{sgn} x) e^{ix2m/w}. \end{cases}$$

Effectively this means that $\{\varphi_n\}$ is already an orthogonal basis. Finally, from Appendix A, nos. 8 and 11, we obtain

$$\sqrt{2\pi}\,(M_1^\star)^\vee(t) = \operatorname{sinc} wt,$$

$$\sqrt{2\pi}\,(M_2^\star)^\vee(t) = -\operatorname{sinc}\left(\frac{wt}{2}\right)\sin\left(\frac{\pi wt}{2}\right),$$

and then a simple step takes us from (12.2.6) to (12.2.8).

It should be observed that the rôles of $\{\varphi_n\}$ and $\{\varphi_n^\star\}$ can be reversed in Theorem 12.8 to obtain a dual series. Of course, when $\{\varphi_n\}$ is an orthonormal basis the sampling series it generates will be self-dual; thus the "function and Hilbert transform" series of Example 12.9 is self-dual.

PROBLEMS

12.1 Show that $Y_\alpha(t) = \operatorname{sinc} w\alpha t \cos \pi wt$ belongs to $PW_{(1+\alpha)w}$, $0 < \alpha < 1$, and is extremal for the aliasing error bound in Example 12.6. (Properties of the Hilbert transform in Appendix B will be needed). (Beaty and Higgins 1994).

12.2 "Function and derivative sampling". Find transfer functions appropriate for reconstructing a Paley–Wiener function f from samples of both itself and its first derivative.

The series, which we met in Problem 1 of Chapter 9, can be written in the form

$$f(t) = \sum_{n\in\mathbb{Z}} f\left(\frac{2n}{w}\right)\operatorname{sinc}^2\frac{w}{2}\left(t-\frac{2n}{w}\right)$$

$$+ \frac{2}{\pi w}f'\left(\frac{2n}{w}\right)\operatorname{sinc}\frac{w}{2}\left(t-\frac{2n}{w}\right)\sin\frac{\pi w}{2}\left(t-\frac{2n}{w}\right).$$

What is the dual series?

12.3 Suppose that p channels consist of p different time shifts in reading samples. Find the appropriate transfer functions and the consequent sampling series (this is called pth order sampling by Kohlenberg (1953)).

12.4 Show that, for every $f \in PW_{\pi w}$, there is the integral analogue

$$f(t) = w\int_{\mathbb{R}} f(2v)\operatorname{sinc}^2\frac{w}{2}(t-2v) + \frac{2}{\pi w}f'(2v)\operatorname{sinc}\frac{w}{2}(t-2v)\sin\frac{\pi w}{2}(t-2v)\,dv$$

of the series in Problem 12.2.

Show that, for every $f \in PW_{\pi w}$, there is the integral analogue

$$f(t) = -w\int_{\mathbb{R}} f^\sim(v)\operatorname{sinc}\frac{w}{2}(t-v)\sin\frac{\pi w}{2}(t-v)\,dv$$

of the series (12.1.5). What about its companion?

12.5 Is there an integral analogue of the series (12.2.8)?

The integral analogues in Problem 12.4 and the possibility suggested in Problem 12.5 could be thought of as reproducing formulae, reproduction occurring through channels rather than from just the function itself.

13

MULTI-BAND SAMPLING

In this chapter we shall discuss sampling theories for functions f such that supp $f^\wedge \subseteq \overline{B}$, where $B \subset \mathbb{R}$ is the union of finitely many disjoint components. We take B to consist of intervals of positive length, each closed on the left and open on the right. Furthermore, these intervals are to be separated by gaps which are themselves intervals of positive length. It will be recalled from Definition 6.19 that such a function is called *multi-band*.

Throughout this and subsequent chapters the set B will be called a *band region*, or if we wish to stress its multi-band character, a *multi-band region*. Its components are usually referred to as *bands*. The distance between the mid-points of any two bands is called the *mid-point band separation*. By PW_B we shall mean the Paley–Wiener space $\{f : f \in L^2(\mathbb{R}) \cap C(\mathbb{R}); \text{ supp } f^\wedge \subseteq \overline{B}\}$.

We shall want our sampling series in the multi-band setting to retain as many amenable features associated with the cardinal series for members of $PW_{\pi w}$ as possible. Chief among these will be the minimal rate, or density, of the sample point distribution, its regularity (regular sampling is convenient for applications[1]), and the orthogonal nature of the expansion. It is inevitable, however, that the reconstruction functions will become more complicated as the band regions take on more geometrical complexity.

In §13.1 we find that all three of these features can be retained when the multi-band region B satisfies a certain geometrical constraint; this case is called *regular sampling*. Next, in §13.2 we find that the regularity of the sampling point distribution can be retained for any multi-band region B, but usually at the expense of incurring a sampling rate above the N–L minimum rate. In §§13.3, 13.4 and 13.5 we look at some special band-region geometries well known from applied fields and find that the N–L minimal sampling rate can be achieved, but the expansions are with respect to a Riesz basis rather than an orthogonal basis, and the sample points have a distribution which is only partially regular (but not completely irregular, as in Chapter 10).

The Dirac delta method for deriving the cardinal series, mentioned in §1.3, is worth bearing in mind throughout the present chapter; particularly that aspect of it in which the closest packing of translated copies of the band-region led us to the minimal sampling rate. We shall use more rigorous methods than the delta method here, but its ideas are nevertheless an invaluable guide to the intuition.

---

[1]This has been standard dogma for many years. However, an eloquent plea for recognition of the advantages, in signal processing applications, of irregular sampling has recently been made by Bilinskis (1995).

The following two properties of sets and their translates will be needed. The second of these is the stronger, and is sometimes referred to as "tiling"; here we shall prefer the more classical *tessellation*.[2] This word referred to two dimensions originally, but there will be no difficulty in using the idea in one dimension, or in higher dimensions in Chapter 14.

Definition 13.1 *A set $B \subset \mathbb{R}$ is said to have disjoint translates by integer multiples of a parameter ρ if*

$$\{B + l\rho\} \cap \{B + m\rho\} = \emptyset, \qquad l, m \in \mathbb{Z}, \ l \neq m. \tag{13.0.1}$$

Definition 13.2 *Let $B \subset \mathbb{R}$, let $\theta_k \in \mathbb{R}$, $k \in \mathbb{Z}$ and let $B_k = B + \theta_k$. The set B is said to tessellate $\mathbb{R}$, or to be a prototile for $\mathbb{R}$, if $\{B_k\}$ is a partition of $\mathbb{R}$; i.e. $\mathbb{R} = \cup_{k \in \mathbb{Z}} B_k$ and $B_k \cap B_l = \emptyset$; $k, l \in \mathbb{Z}$, $k \neq l$.*

13.1   Regular sampling

Let the interval $I = [-\pi w, \pi w)$ be given. We shall construct a multi-band region B as follows. Let I be partitioned into a finite number, K say, of half-closed, half-open sub-intervals I_j, $j = 1, \ldots K$, such that $I = \cup_j I_j$, $I_m \cap I_n = \emptyset$, $m, n = 1, \ldots, K$. Now let the bands of B consist of intervals which are obtained by translating each of the I_j's along $\mathbb{R}$ through some integer multiple of $2\pi w$. A good way to visualize this construction is to think of a wheel whose circumference is of length $2\pi w$. It is rolled accross the interval I, and as this happens each sub-interval I_j is marked on its circumference. The wheel now rolls along the time axis, and after a certain number of revolutions one of the I_j's is "peeled off" the wheel and deposited on the axis; likewise, after further revolutions the other I_j's are peeled off and each is deposited somewhere, once and once only. Then B consists of the union of these depositions.

An important property of B is that its translates by integer multiples of $2\pi w$ tessellate $\mathbb{R}$, as the reader can easily verify.

A multi-band region of this kind is sometimes called an *explosion* of I. It has been described rather intuitively; a properly mathematical description is given in §14.3 where it is called a *fundamental region*.

It is not difficult to give a direct proof of the following theorem on regular sampling; however, this theorem is the one-dimensional case of a multi-dimensional regular sampling theorem, to be proved in §14.3, so we shall not give a proof in the text here.

Theorem 13.3 (Regular Sampling Theorem) *Let B be an explosion of $I = [-\pi w, \pi w)$. Then, for every $f \in PW_B$,*

---

[2]From the Latin *tessellare*, to form a mosaic, especially a mosaic pavement. The Latin noun is *tessella*, which is a diminutive form of *tessera*. Going back to Attic Greek times the word *tessera* (literally "four") meant a small, usually cubical piece of marble, tile, etc., used for making mosaic pavements; it subsequently passed into Latin and thence to English with this meaning.

$$f(t) = \sum_{n \in \mathbb{Z}} f\left(\frac{n}{w}\right) g\left(t - \frac{n}{w}\right),$$

where

$$g(t) = \frac{1}{2\pi w} \int_B e^{ixt}\, dx = \frac{1}{\sqrt{2\pi w}} \chi_B^{\vee}(t).$$

The set $\{g(t - n/w)\}$, $n \in \mathbb{Z}$ is an orthonormal basis for PW_B and the corresponding Parseval relation is

$$\sum_{n \in \mathbb{Z}} \left| f\left(\frac{n}{w}\right)\right|^2 = w\|f\|^2.$$

Note that the sampling rate is w samples per second and that this equals $m(B)/2\pi$, the N–L minimum sampling rate.

13.2   Optimal regular sampling

Suppose we are given a multi-band region B. If our task were simply to find a sampling series for members f of PW_B we could enclose B in an interval I and expand f in the ordinary cardinal series for $f \in PW_I$. This procedure certainly has its merits; the expansion is an orthonormal one, the reconstruction functions are sinc functions and so are very simple, and the sample points are regularly distributed. The disadvantage is that the sampling rate may be well above the N–L minimum if there are large gaps in B, as is evident by comparing $m(B)$ with the larger $m(I)$. We shall see that it may be possible to exploit gaps in B by enclosing it in an explosion of a smaller interval, one which is perhaps much closer in length to $m(B)$ than to $m(I)$ (or even equal in length if a smaller interval can be found that explodes to B). When the smallest such interval is determined we have reached an "optimal" sampling rate much nearer to the N–L minimum than could be obtained by enclosing B in I. Furthermore, the pleasant features mentioned above are retained, in particular the regularity of the sample point distribution.

This process is called *optimal regular sampling*. The concept is due to Dodson and Silva (1989), upon whose work the present section is based.

In this section it will be convenient to speak of a *sampling set* $\{\lambda_n\}$ for PW_B when there is some process whereby every $f \in PW_B$ can be reconstructed from its samples $\{f(\lambda_n)\}$.

Theorem 13.4 (Optimal Regular Sampling) *Let B be a multi-band region, and let τ be a positive real number such that translates of B through all integer multiples of $2\pi\tau$ are disjoint. Then $(1/\tau)\mathbb{Z}$ is a sampling set for $f \in PW_B$, and we have*

$$f(t) = \sum_{n \in \mathbb{Z}} f\left(\frac{n}{\tau}\right) g\left(t - \frac{n}{\tau}\right),$$

where $g(t) = \chi_B^{\vee}(t)/\tau\sqrt{2\pi}$. The convergence is globally uniform over $\mathbb{R}$.

Proof The interval $J = [-\pi\tau, \pi\tau)$ tessellates $\mathbb{R}$ by translation through integer multiples of $2\pi\tau$, and by hypothesis B has disjoint translates by the same set of translations. It therefore must be the case that either B is an explosion of J, or that there is at least one sub-interval of J which does not meet B or any of its translates. All sub-intervals of J of this kind can now be adjoined to B to form an augmented set $\tilde{B}$, which is of course an explosion of J. Alternatively, any sub-interval of J of this kind can be subjected to translation through an integer multiple of $2\pi\tau$ before being incorporated into $\tilde{B}$, so $\tilde{B}$ is not unique; but in any case it is always an explosion of J.

Now since $B \subseteq \tilde{B}$, the Regular Sampling Theorem 13.3 with band region $\tilde{B}$ applies to $f \in PW_B$. This gives the required expansion, in which the reconstruction function g can be calculated with respect to B, because $f^\wedge$ vanishes outside B whenever $f \in PW_B$.

The uniformity follows from the Convergence Principle of §6.6.

Usually one seeks the *smallest* value of τ so that the construction in the theorem holds. This is easy to calculate directly in simple cases; Corollary 13.6 below is an example. Fortunately, there is a simple and ingenious algorithm for doing this calculation in all cases (it will be found in the next section). Meanwhile we shall look at a special case.

Definition 13.5 When $B = [-\pi(w_0 + w), -\pi w_0) \cup [\pi w_0, \pi(w_0 + w))$, PW_B is called the class of band-pass functions.

Corollary 13.6 (Optimal Regular Sampling; Band-pass Case) Let B be as in Definition 13.5. Then $(1/\tau_0)\mathbb{Z}$ is an optimal regular sampling set for $f \in PW_B$, where

$$\tau_0 = (w_0 + w) / \left[\frac{w_0 + w}{w}\right]. \qquad (13.2.1)$$

Here, and in the proof to follow, [ ] means as usual the "integer part". The reconstruction function g, defined in Theorem 13.4, is given by

$$g(t) = \frac{w}{\tau_0} \operatorname{sinc} \frac{wt}{2} \cos \frac{\pi}{2}(2w_0 + w)t.$$

Proof Let τ be any real positive number such that translates of B through all integer multiples of $2\pi\tau$ are disjoint. Let $r \in \mathbb{N}$ be such that the rth translate through $2\pi\tau$ (to the right) of B is the first that is positioned wholly to the right of B. These facts are expressed by the two inequalities

$$2\pi\tau r \geq 2\pi(w_0 + w) \qquad \text{and} \qquad 2\pi\tau(r - 1) \leq 2\pi w_0.$$

On combining these we obtain $r \leq (w_0 + w)/w$, so the largest value that r can take is $[(w_0 + w)/w]$. When this is substituted into the first inequality we find the smallest value that τ can take; it is τ_0, given by (13.2.1). The form of g follows from Appendix A, no. 16.

It is interesting to note that if w_0 is an integer multiple of w we find that the optimal regular sampling rate τ_0 is equal to w, the N–L minimal sampling rate. The minimal sampling rate is achieved in this case; otherwise of course it is not, but in any case τ_0 and w do not differ greatly.

13.3   An algorithm for the optimal regular sampling rate

It will be convenient to describe the algorithm in terms of difference sets.

Definition 13.7 *Given a set $B \subset \mathbb{R}$, its difference set $D(B)$ is defined by*

$$D(B) := \{x - y : \ x, y \in B\}.$$

Clearly, $D(B)$ is a symmetric set; that is, $x \in D(B)$ implies $-x \in D(B)$.

Lemma 13.8 *The disjoint translates condition (13.0.1) is equivalent to the difference set condition*

$$D(B) \cap \rho\mathbb{Z} = \{0\}. \tag{13.3.1}$$

Proof Suppose the disjoint translates condition (13.0.1) holds. Then for $x, y \in B$ and $l, m \in \mathbb{Z}$, $l \neq m$, we have $x + l\rho \neq y + m\rho$, or $x - y \neq (m - l)\rho$. But $0 \in D(B)$, so (13.3.1) holds.

On the other hand, suppose (13.3.1) holds. Then $k\rho \notin D(B)$ when $k \neq 0$; that is, $x - y \neq k\rho$ for any $x, y \in B$. Therefore $x + l\rho \neq y + (k + l)\rho$, $l \in \mathbb{Z}$, and so (13.0.1) holds.

The following is an algorithm for finding the least value of ρ for disjoint translates sampling.

Let $B = \cup_{i=1}^{M}(c_i, d_i)$. Then $D(B)$ is also a union of intervals, say $D(B) = \cup_{i=1}^{N}(C_i, D_i)$. We wish to find the least value of ρ such that B has disjoint translates by integer multiples of ρ; or, by Lemma 13.8, the least value of ρ such that (13.3.1) holds. Because of symmetry we need only consider the positive half of $D(B)$.

The process is an algorithmic one. We start by trying $\rho_1 = m(B)$; clearly, no smaller starting value is possible. If $D(B) \cap \rho_1\mathbb{Z} = 0$ the process terminates and ρ_1 is the optimal sampling rate. If, on the other hand, an integer multiple of ρ_1 belongs to $D(B)$, let k be the least integer such that $k\rho_1 \in (C_r, D_r)$, for some r, $1 \leq r \leq N$. Then augment ρ_1 by putting

$$\rho_2 = \frac{D_r}{k} = \rho_1 + \frac{D_r - k\rho_1}{k} > \rho_1.$$

Let us call this "augmentation at D_r". Again, the process terminates if $D(B) \cap \rho_2\mathbb{Z} = 0$, and ρ_2 is optimal.

It is now clear that if the process does not terminate at a particular stage, then the algorithm is continued by a further augmentation of ρ at some right end point $D_{r'}$, $1 \leq r' \leq N$, of a component of $D(B)$.

It remains to show that the algorithm always terminates in a finite number of steps, yielding the optimal value of ρ. To see this, it is important to note that at each augmentation at D_r of, say, ρ_s to ρ_{s+1}, the number of integer multiples of ρ_{s+1} which are strictly less than D_r (that is, the number of "hops" of length ρ_{s+1} occurring strictly to the left of D_r) is one less than the number of hops of length ρ_s occurring strictly to the left of D_r. Thus the algorithm will eventually reduce to zero the number of hops whose length is the current value of ρ occurring strictly to the left of D_r, unless termination occurs first.

In the absence of termination, the algorithm will continue to have the same reducing effect at every D_r, $1 \leq r \leq N$, termination finally occurring with the optimal value $\rho = D_N$. In any case termination occurs in finitely many steps, and the algorithm is complete.

The simplest example is when $N = 1$ so that B is a single interval; then the optimal value $\rho = D_1 = 2d_1$ occurs at the first iteration. When $B = (-d_1, -d_1/2) \cup (d_1/2, d_1)$,[3] termination also occurs at the first iteration, this time with the optimal value $\rho = d_1$.

13.4   Selectively tiled band regions

Up to this point we have sampled and reconstructed multi-band functions from samples at points whose distribution has been regular. The N–L sampling rate could always be achieved if B was an explosion of an interval, but if this was not the case, then we had to settle for the optimal regular sampling rate which was not necessarily N–L.

Suppose we now insist that our sampling rate be the N–L minimum. This raises the interesting question of what band regions B are then allowable, and what is the nature of the sample point distribution. This is a specialist topic and is planned for inclusion in H–S. We cannot enter into all the details here; however, in this section and the next we shall look at two of the simpler possibilities.

Probably the simplest type of multi-band region B that is not necessarily an explosion of an interval, but which allows sampling and reconstruction for $f \in PW_B$ at the N–L rate, can be described as follows. Take an interval I and subdivide it into a certain number of equal sub-intervals; then select some of these sub-intervals to make up B. This means that the bands of B have lengths that are all integer multiples of some basic length, and the same is true of the gaps between them. To put the matter succinctly, the bands and the gaps are *commensurable*. Such a B will be called a *selectively tiled band-region*. The basic study in this area is by Bezuglaya and Katsnelson (1993); the present section

---

[3]The "polaroid camera example". On a polaroid camera film, two exposures per picture are required, and they are separated so that one pair interlaces with the previous pair according to the same scheme whereby $B = (-d_1, -d_1/2) \cup (d_1/2, d_1)$ tessellates $\mathbb{R}$ by translation.

is based on a modification of their treatment, in which ideas of multi-channel sampling are incorporated (Higgins (to appear)).

We seek a sampling series for $f \in PW_B$ of the form

$$f(t) = \sum_{n \in \mathbb{Z}} (\mathcal{S}f)(\lambda_n) R_n(t),$$

where $\mathcal{S}$ is a transformation defined on PW_B, $\Lambda := (\lambda_n)$ is a real sequence, and $\{R_n\}$ is a set of reconstruction functions. When $\mathcal{S}$ is a single transformation we refer, as in §12.1, to *single channel sampling*; when $\mathcal{S}$ is a sum of several transformations we refer to *multi-channel sampling*.

First suppose that $B = \cup_1^N B_j$, where B_j are intervals, all of length d where d is some positive constant, and that the gaps between successive B_j's are integer (or zero) multiples of d. Now let $\mathcal{T} : L^2(B_1) \oplus \cdots \oplus L^2(B_N) \mapsto L^2(B)$ be given by

$$(g_1, \ldots, g_N) \mapsto M_1(x)(g_1(x))_d + \cdots + M_N(x)(g_N(x))_d,$$

where $g_j \in L^2(B_j)$, $(\ )_d$ means periodic extension with period d, and each $M_j(x)$ is a bounded measurable function defined on B. Now

$$w_{Nn}(x) = (e^{2\pi i n x/d}, \theta, \ldots, \theta)$$

$$w_{Nn+1}(x) = (\theta, e^{2\pi i n x/d}, \ldots, \theta)$$

$$\vdots$$

$$w_{Nn+N-1}(x) = (\theta, \ldots, \theta, e^{2\pi i n x/d})$$

(in which θ denotes the null function, and n ranges over $\mathbb{Z}$) forms an orthogonal basis for $L^2(B_1) \oplus \cdots \oplus L^2(B_N)$.

It is now easily seen that

$$(\mathcal{T}w_{Nn+j})(x) = M_j(x)e^{2\pi i n x/d}, \qquad j = 1, \ldots, N.$$

We wish to show that $\{\mathcal{T}w_{Nn+j}\}$, $j = 1, \ldots, N$, $n \in \mathbb{Z}$, forms a Riesz basis for $L^2(B)$. For this purpose let us first note that in defining the direct sum above we needed to decompose members of $L^2(B)$ into components. In order to compare these two Hilbert spaces, it is convenient to make another, slightly different, decomposition. Thus, let us put

$$G = G_1 + \cdots + G_N \in L^2(B), \qquad G_j \in L^2(B_j), \ j = 1, \ldots, N.$$

This does not change $L^2(B)$, but it does make it possible to specify G by letting its argument range over B_1 only. Thus

$$G(x) = G_1(x) + G_2(x + l_2 d) + \cdots + G_N(x + l_N d), \qquad x \in B_1,$$

where l_j is the mid-point band separation of B_1 and B_j. We need to know whether there exist g_j s such that

$$G_j = M_1(x)g_1(x) + \ldots + M_N(x)g_N(x),$$

for $x \in B_j$ (where for convenience the subscript d indicating periodic extension has been omitted); or, equivalently,

$$G_j(x + l_jd) = \sum_{n=1}^{N} M_n(x + l_jd)g_n(x), \qquad j = 1, \ldots, N$$

for $x \in B_1$. These equations can be written

$$
\begin{pmatrix}
G_1(x) \\
G_2(x + l_2d) \\
\vdots \\
G_N(x + l_Nd)
\end{pmatrix}
$$

$$
= \begin{pmatrix}
M_1(x) & M_2(x) & \cdots & M_N(x) \\
M_1(x + l_2d) & M_2(x + l_2d) & \cdots & M_N(x + l_2d) \\
\vdots & \vdots & & \vdots \\
M_1(x + l_Nd) & M_2(x + l_Nd) & \cdots & M_N(x + l_Nd)
\end{pmatrix}
\begin{pmatrix}
g_1(x) \\
g_2(x) \\
\vdots \\
g_N(x)
\end{pmatrix}.
$$

Let the matrix of coefficients here be denoted by Δ; then if $\det \Delta \neq 0$ the g_j's are defined in terms of given G_j's for $x \in B_1$ and hence by periodicity for all $x \in B$. Indeed when $\det \Delta \neq 0$, $\mathcal{T}$ is one-to-one and onto. Clearly $\mathcal{T}$ is linear, so it remains to show that it is bounded.

Let $\sup_{x \in B} M_j(x) := \mu_j$. Then

$$
\begin{aligned}
\|M_1g_1 + \cdots + M_Ng_N\|_B^2 &\leq \{\|M_1g_1\|_B + \cdots + \|M_Ng_N\|_B\}^2 \\
&\leq 2^{N-1}\{\|M_1g_1\|_B^2 + \cdots + \|M_Ng_N\|_B^2\} \\
&\leq 2^{N-1}N\{\mu_1^2\|g_1\|_{B_1}^2 + \cdots + \mu_N^2\|g_N\|_{B_N}^2\} \\
&\leq 2^{N-1}N \max_{j=1,\ldots,N} \mu_j^2\{\|g_1\|_{B_1}^2 + \cdots + \|g_N\|_{B_N}^2\} \\
&= (\text{constant})\|(g_1, \ldots, g_N)\|^2.
\end{aligned}
$$

This proves that $\{\mathcal{T}w_{Nn+j}\}$, $j = 1, \ldots, N$, $n \in \mathbb{Z}$, forms a Riesz basis for $L^2(B)$ as required.

$$* \qquad * \qquad *$$

With the notations already established, suppose that $\det \Delta \neq 0$, so that $\{\mathcal{T}w_{Nn+j}\}$, $j = 1, \ldots, N$, $n \in \mathbb{Z}$, forms a Riesz basis for $L^2(B)$. Let us denote by $\{\psi_{n,j}\}$, $j = 1, \ldots, N$, $n \in \mathbb{Z}$, the associated biorthonormal basis; as we know this is also a Riesz basis. Let $\mathcal{S}_j := \mathcal{F}^{-1} M_j \mathcal{F}$.

Theorem 13.9 *Let B be a selectively tiled band-region, and let $f \in PW_B$. Then the sampling series representation*

$$f(t) = \sqrt{2\pi} \sum_{n \in \mathbb{Z}} \sum_{j=1}^{N} (\mathcal{S}_j f) \left(\frac{2n\pi}{d} \right) (\overline{\psi}_{n,j})^{\vee} (t)$$

holds, with convergence in the norm of PW_B, and also pointwise and globally uniformly over $\mathbb{R}$.

Proof Because of the Riesz basis properties already established, we have in the norm of $L^2(B)$

$$f^{\wedge}(x) = \sum_{n \in \mathbb{Z}} \sum_{j=1}^{N} c_{n,j} \overline{\psi}_{n,j}(x),$$

where

$$c_{n,j} = \langle f^{\wedge}, \overline{m}_j(\cdot) e^{-2\pi i n \cdot / d} \rangle_{L^2(B)}$$
$$= \sqrt{2\pi} \left(\mathcal{F}^{-1} M_j \mathcal{F} \right) \left(\frac{2n\pi}{d} \right) = \sqrt{2\pi} \, (\mathcal{S}_j f) \left(\frac{2n\pi}{d} \right).$$

The standard Fourier duality now gives the required result in the norm of PW_B.

The uniform convergence follows from the Convergence Principle of §6.6. The reader is invited to formulate an associated stability result in Problem 13.11.

Example 13.10 Choose $M_j(x) := e^{ik_j x}$, $j = 1, \ldots, N$, where the k_j are distinct real numbers so chosen that

$$\det \Delta = e^{ik_1 x} \ldots e^{ik_N x} \begin{vmatrix} 1 & \cdots & 1 \\ e^{ik_1 l_2 d} & \cdots & e^{ik_N l_2 d} \\ \vdots & & \vdots \\ e^{ik_1 l_N d} & \cdots & e^{ik_N l_N d} \end{vmatrix}$$

does not vanish. Then $(\mathcal{S}_j f)(x) := f(x + k_j)$.

Bezuglaya and Katsnelson (1993) take $k_j = j\pi / l_N d$, $j = 1, \ldots, N$. Then the previous determinant becomes

$$\begin{vmatrix} 1 & \cdots & 1 \\ e^{\pi i l_2 / l_N} & \cdots & e^{N\pi i l_2 / l_N} \\ \vdots & & \vdots \\ e^{\pi i l_N / l_N} & \cdots & e^{N\pi i l_N / l_N} \end{vmatrix},$$

a Vandermonde determinant having the value

$$\prod_{q=2}^{N}\left\{1 - e^{\pi i l_q/l_N}\right\} \prod_{2 \le r < s \le N}\left\{e^{\pi i l_r/l_N} - e^{\pi i l_s/l_N}\right\} \ne 0.$$

In this last example sampling was through channels whose effects were to time shift the object function. Equivalently, we might have simply sampled the function itself at shifted sample points; this is the more common description to be found in the literature on multi-band sampling.

Theorem 13.9 is really just an existence theorem for representing selectively tiled band-limited functions in sampling series. In fact, calculation of the reconstruction functions $\{(\overline{\psi}_{n,j})^\vee(t)\}$ can be difficult in practice (however, it was done by Bezuglaya and Katsnelson for their special case of Example 13.10 above; the results will be found in Problem 13.5). The sample point spacing in the series is $2\pi/d$. Since N samples are taken at each point, the sampling rate is $Nd/2\pi = m(B)/2\pi$ samples per second, and this is the N–L rate. It is not known whether this rate is also minimal for multi-channel sampling.

13.5   Harmonic signals

Suppose that a multi-band function is such that its bands are of equal length, and the gaps between these bands are also all of equal length (there is no requirement of commensurability in general). Beaty and Dodson (1993) use the name *harmonic signal* for this kind of function, and point out that they " ... arise in practice, typically when information is transmitted using a carrier frequency". Beaty and Dodson give a complete discussion of these functions and their allowable sampling rates with masterful elegance, but it is well beyond our present scope.

The purpose of this section is to show that when the band region of a harmonic signal is of the selectively tiled type (so that we do have commensurability between bands and gaps), and if certain time shifts are used for the channels, the sampling series of Theorem 13.9 is an orthogonal expansion rather than the more general Riesz basis expansion, and the reconstruction functions are of a standard type. We hope to capture once again the spirit of Hardy's classic treatment of the ordinary cardinal series (which was attempted in Lemma 6.20).

Details are given for the case $K = 2N$. The case $K = 2N + 1$ is very similar.

Lemma 13.11 *Let $B = \cup_{j=1}^{K} B_j$ be a selectively tiled band-region in which B_j is an interval of length πw for every j, and the gap separating B_j from B_{j+1} is of length $\rho \pi w$ for some fixed positive integer ρ. Then the set*

$$\varphi_{n,\alpha}(x) = e^{2inx/w}\, e^{i\alpha x}, \qquad \alpha = 0, \frac{1}{N\rho w},\dots,\frac{2N-1}{N\rho w}, \quad n \in \mathbb{Z}$$

is orthogonal and complete in $L^2(B)$.

Proof   The set is already a Riesz basis, by the analysis preceding Theorem 13.9 (Δ is clearly non-singular), so we need only look at the orthogonality. Now

$$\int_B \varphi_{m,\alpha}(x)\, \overline{\varphi}_{n,\beta}(x)\, dx = \sum_{j=1}^{2N} \int_{B_j} e^{2i(m-n)x/w}\, e^{i(\alpha-\beta)x}\, dx$$

$$= \sum_{j=1}^{2N} \left\{ e^{[2(m-n)+(\alpha-\beta)w]i(j-1)\pi\rho} \right\} \int_{B_1} \varphi_{m,\alpha}(x)\, \overline{\varphi}_{n,\beta}(x)\, dx.$$

This expression vanishes, because the integral vanishes if $\alpha = \beta$, $m \neq n$; and if $\alpha \neq \beta$ the GP sums to

$$\frac{1 - e^{2iN\pi\rho[2(m-n)+(\alpha-\beta)w]}}{1 - e^{i\pi\rho[2(m-n)+(\alpha-\beta)w]}},$$

and this vanishes for every m and n.

The sampling series using this orthonormal basis now follows from Theorem 13.9. It is

$$f(t) = \sqrt{2\pi} \sum_{n \in \mathbb{Z}} \sum_{\alpha} f\left(\frac{2n}{w} - \alpha\right) \chi_B^{\vee}\left(t - \frac{2n}{w} - \alpha\right), \qquad (13.5.1)$$

for a harmonic signal $f \in PW_B$. The form of the reconstruction functions follows from standard Fourier transform formulae.

13.6   Band-pass sampling

We take up the band-pass case already introduced in Definition 13.5, this time with the requirement of achieving the N–L sampling rate. Throughout this section we take $B = [-\pi(w_0 + w), -\pi w_0) \cup [\pi w_0, \pi(w_0 + w))$, the "band-pass" band-region. Note that $m(B) = 2\pi w$; hence the N–L sampling rate is w samples per second.

As in the previous section, the number of channels must match the number of frequency bands; that is, two. Each channel produces samples at half the N–L rate, so that each channel is weighted equally in this respect, just as in the selectively tiled case. In this section one of the channels will be chosen, for convenience, to be the "identity", so the samples associated with this channel are those of the function itself.

In Example 13.14 below, the "other channel" is chosen to be the Hilbert transform, giving one of the first band-pass sampling theorems to appear in the literature, that of Goldman (1953, p. 76) and Woodward (1953, p. 34). The Hilbert transform appeared quite naturally in the derivation, but is not particularly convenient for applications. In Problem 13.6, the other channel is chosen to be a (nearly arbitrary) time shift, and we obtain the well known sampling series of Kohlenberg (1953). While this may be more convenient in some respects, it does require surprisingly complicated reconstruction functions.

Here we are going to collect these special cases under one umbrella, the Riesz basis method (Kohlenberg's proof used the delta method). The establishment of an appropriate Riesz basis property is harder in the band-pass case than it was in the selectively tiled case, even though the number of bands is only two. This is because the gap between the two components is not now commensurable with the component length. A refinement of the previous method will be necessary.

Let B be the band-pass band-region mentioned at the beginning of this section, and let $r = [2w_0/w] + 1$, where as usual [ ] denotes the "integer part".

As usual, let $M(x)$ be a bounded measurable function defined on B, and let $\mathcal{M}$ denote the operation of multiplication by M.

Let us put $B^- = [-\pi(w_0 + w), -\pi w_0]$, and $B^+ = [\pi w_0, \pi(w_0 + w)]$. Let $L^2(B^-) \oplus L^2(B^+)$ denote the direct sum consisting of pairs (g^-, g^+) with $g^- \in L^2(B^-)$ and $g^+ \in L^2(B^+)$. Further, let $\mathcal{Q}$ denote the mapping

$$L^2(B^-) \oplus L^2(B^+) \longmapsto L^2(B)$$

given by

$$\mathcal{Q}: (g^-(x), g^+(x)) \longmapsto F(x) = \left(g^-(x)\right)_p + M(x)\left(g^+(x)\right)_p, \qquad x \in B,$$

where subscript p denotes periodic extension with period πw. Next, the set $\{w_n\}$, $n \in \mathbb{Z}$, of (12.2.1) is also an orthonormal basis for $L^2(B^-) \oplus L^2(B^+)$. In particular, since the exponentials are already periodic with period πw,

$$\left. \begin{array}{ll} \mathcal{Q}: w_{2m} \longmapsto e^{ix2m/w} & = \varphi_{2m}(x) \\ \mathcal{Q}: w_{2m+1} \longmapsto M(x)e^{ix2m/w} & = \varphi_{2m+1}(x) \end{array} \right\}, \qquad x \in B. \qquad (13.6.1)$$

Lemma 13.12 *The set (13.6.1) is a Riesz basis for $L^2(B)$.*

Proof Now it is the rth translate through πw to the right of B^- which first overlaps B^+; this region of overlap is denoted by R_1^+, and that part of B^+ which overlaps B^- at the rth translate to the left is denoted by R_1^-. Similarly, R_2^+ and R_2^- play the same rôles *vis-à-vis* the $(r+1)$th translates.

Let us now put

$$\begin{array}{ll} F_1^+ = F \mid R_1^+, & g_1^+ = g^+ \mid R_1^+, \\ F_2^+ = F \mid R_2^+, & g_2^+ = g^+ \mid R_2^+, \\ F_1^- = F \mid R_1^-, & g_1^- = g^- \mid R_1^-, \\ F_2^- = F \mid R_2^-, & g_2^- = g^- \mid R_2^-. \end{array}$$

We need to identify F in terms of g^- and g^+. Thus, by periodicity we have

$$\text{for } x \in R_1^+: \quad F_1^+(x) = g_1^-(x) + g_1^+(x)M(x), \qquad (13.6.2)$$
$$\text{for } x \in R_2^+: \quad F_2^+(x) = g_2^-(x) + g_2^+(x)M(x), \qquad (13.6.3)$$
$$\text{for } x \in R_1^-: \quad F_1^-(x) = g_1^-(x) + g_1^+(x)M(x), \qquad (13.6.4)$$
$$\text{for } x \in R_2^-: \quad F_2^-(x) = g_2^-(x) + g_2^+(x)M(x). \qquad (13.6.5)$$

Let us re-write (13.6.4) and (13.6.5) as follows:

for $x \in R_1^+ :$ $F_1^-(x - r\pi w) = g_1^-(x) + g_1^+(x)M(x - r\pi w);$ (13.6.6)

for $x \in R_2^+ :$ $F_2^-(x - (r+1)\pi w) = g_2^-(x) + g_2^+(x)M(x - (r+1)\pi w).$ (13.6.7)

Taken together, (13.6.2) and (13.6.6) give

$$\begin{pmatrix} F_1^+(x) \\ F_1^-(x - r\pi w) \end{pmatrix} = \begin{pmatrix} 1 & M(x) \\ 1 & M(x - r\pi w) \end{pmatrix} \begin{pmatrix} g_1^-(x) \\ g_1^+(x) \end{pmatrix}.$$ (13.6.8)

Now whenever the determinant

$$\begin{vmatrix} 1 & M(x) \\ 1 & M(x - r\pi w) \end{vmatrix}$$ (13.6.9)

does not vanish, (13.6.8) has a unique solution. A similar determinant is obtained (but with r replaced by $(r+1)$), if we take (13.6.3) together with (13.6.7). These determinants do not vanish if the criteria

$$\left. \begin{array}{ll} M(x) \neq M(x - r\pi w), & \text{for } x \in R_1^+ \\[2mm] M(x) \neq M(x - (r+1)\pi w), & \text{for } x \in R_2^+ \end{array} \right\}$$ (13.6.10)

hold, and then we find that the g's and the F's determine each other uniquely, so that Q is then one-to-one and onto.

Furthermore, Q is bounded; the proof is much the same as that for the boundedness of T in §13.4. We now see that Q is an isomorphism. Hence the φ's on the right in (13.6.1) form a set that is fully equivalent to $\{w_n\}$; therefore it is a Riesz basis for $L^2(B)$.

From this Riesz basis property we can now obtain a sampling series for band-pass functions.

Theorem 13.13 Let $B = [-\pi(w_0 + w), -\pi w_0) \cup [\pi w_0, \pi(w_0 + w))$ and let $f \in PW_B$. Let M satisfy (13.6.10), and let

$$g(t) = \frac{1}{\sqrt{2\pi}} \int_B M(x)f^\wedge(x)e^{ixt} \, dx.$$

Let $\{\varphi_n\}$ be the Riesz basis of (13.6.1), and let its associated biorthonormal basis be denoted by $\{\varphi_n^*\}$. Then

$$f(t) = \sum_{n \in \mathbb{Z}} f\left(\frac{2n}{w}\right) R_n(t) + g\left(\frac{2n}{w}\right) S_n(t),$$

where

$$R_n = \left(\varphi_{2n}^*\right)^\vee \quad \text{and} \quad S_n = \left(\varphi_{2n+1}^*\right)^\vee, \qquad n \in \mathbb{Z}.$$

This is a globally uniformly convergent stable sampling series.

Proof The proof follows closely that of Theorem 12.8, part 3.

The convergence follows from the convergence principle of §6.6, and the stability comes from Problem 13.11.

Example 13.14 We look at function and Hilbert transform sampling, as we did in Example 12.9, but this time in the band-pass case.

In Theorem 13.13 take $M(x) = -i\,\mathrm{sgn}\,x$. We find that $\varphi_{2m}(x) = e^{ix2m/w}$ and $\varphi_{2m+1}(x) = -i\,\mathrm{sgn}\,x\,e^{ix2m/w}$, $m \in \mathbb{Z}$. It is then an easy matter of experimentation and verification to find that $(\varphi_n^\star)$ are biorthonormal Riesz bases for $L^2(B)$, where

$$\left.\begin{aligned}
\varphi_{2m}^\star &= \frac{1}{2}\left\{\chi_{B^-}(x) + \chi_{B^+}(x)\right\}e^{ix2m/w}, \\[2mm]
\varphi_{2m+1}^\star &= \frac{i}{2}\left\{\chi_{B^-}(x) - \chi_{B^+}(x)\right\}e^{ix2m/w}.
\end{aligned}\right\}$$

Then we find from Theorem 13.13, using special transforms from Appendix A, that

$$f(t) = \sum_{n\in\mathbb{Z}}\left\{f\left(\frac{2n}{w}\right)\cos w_1\left(t - \frac{2n}{w}\right)\right.$$
$$\left. - f^\sim\left(\frac{2n}{w}\right)\sin w_1\left(t - \frac{2n}{w}\right)\right\}\mathrm{sinc}\,\frac{w}{2}\left(t - \frac{2n}{w}\right), \quad (13.6.11)$$

for $f \in PW_B$, where $w_1 = \frac{\pi}{2}(2w_0 + w)$.

PROBLEMS

13.1 Apply the algorithm of §13.3 in the cases:
(a) $B = (-22, -20) \cup (-6, -3) \cup (3, 6) \cup (20, 22)$;
(b) $B = (-27, -25) \cup (-5, -3) \cup (3, 5) \cup (25, 27)$.
(In (a) the algorithm succeeds at the fifth iteration; in (b) at the fourth.)

13.2 Discuss the general two-band case $B = (a, b) \cup (c, d)$. (There is no single general formula for the optimal sampling rate; see Dodson and Silva (1989). Seip has proved (1995, p. 172) that $L^2((a, b) \cup (c, d))$ has a Riesz basis of complex exponentials; this together with Theorem 10.7 shows that the N–L rate can be achieved in this case).

13.3 Develop a direct approach to the sampling of harmonic signals by iteration of the Fourier transform formula 1 of Appendix A.

13.4 Let a band-region B_1 consist of three intervals; let them all be of length b and be separated by gaps of length b. Also, let B_2 be the same as B_1 except that the gaps are now to be of length $2b$. Develop sampling theories for PW_{B_1} and PW_{B_2}. What is the essential difference between them? Can you generalize?

13.5 Suppose that in the special case of Bezuglaya and Katsnelson in Example 13.10 the sample points $\{\lambda_n\}$ are given by

$$t_{Nm+j-1} = \frac{2m}{w} + (j-1)\alpha, \qquad j = 1, \ldots, N, \ m \in \mathbb{Z}.$$

Show that, giving n a second parametrization $n = Ns + r - 1$, $r = 1, \ldots, N$, $s \in \mathbb{Z}$, the reconstruction functions are of the form

$$S_n(t) = R_{(s,r)}(t) = \frac{e^{-i\lambda_1 2s/w}}{D} D_r(t) \operatorname{sinc} \frac{w}{2}\left(t - \frac{2s}{w} - r\alpha\right), \qquad r = 1, \ldots, N, \ s \in \mathbb{Z},$$

where

$$D = \begin{vmatrix} 1 & e^{i\lambda_1\alpha} & e^{i\lambda_1 2\alpha} & \ldots & e^{i\lambda_1(N-1)\alpha} \\ 1 & e^{i\lambda_2\alpha} & e^{i\lambda_2 2\alpha} & \ldots & e^{i\lambda_2(N-1)\alpha} \\ \vdots & & & & \\ 1 & e^{i\lambda_N\alpha} & e^{i\lambda_N 2\alpha} & \ldots & e^{i\lambda_N(N-1)\alpha} \end{vmatrix}$$

and $D_r(t)$ is obtained from the determinant D by replacing its rth column with the column

$$\begin{pmatrix} e^{i\lambda_1 t} \\ e^{i\lambda_2 t} \\ \vdots \\ e^{i\lambda_N t} \end{pmatrix}.$$

13.6 Show that for every band-pass function f the representation

$$f(t) = \sum_{n=-\infty}^{\infty} f\left(\frac{2n}{w}\right) s\left(t - \frac{2n}{w}\right) + f\left(\frac{2n}{w} + k\right) s\left(-t + \frac{2n}{w} + k\right),$$

$$k \neq \frac{2m}{wr}, \quad k \neq \frac{2m}{w(r+1)}, \quad m \in \mathbb{Z},$$

holds, with convergence as in the Convergence Principle of §6.6, where $r := [2w_0/w] + 1$, and where

$$s(t) = A(w_0 + w, r+1, t) - A(rw - w_0, r+1, t) + A(rw - w_0, r, t) - A(w_0, r, t)$$

with

$$A(W, R, t) := \frac{\cos \frac{1}{2}\pi[2Wt - Rwk]}{\pi wt \sin \frac{1}{2}\pi Rwk}.$$

(Kohlenberg 1953).

13.7 It is natural to enquire whether there are conditions under which the biorthogonal expansion of the previous problem reduces to an orthogonal expansion. Show that appropriate conditions are that $k = 1/wr$ and that $2w_0/w = r - 1$. Show that when and only when the band region satisfies these special conditions,

$$f(t) = \sqrt{2\pi} \sum_{n=0,1\,(2r)} f\left(\frac{n}{rw}\right) \operatorname{sinc}\tfrac{1}{2}\left(wt - \frac{n}{r}\right) \cos\frac{\pi r}{2}\left(wt - \frac{n}{r}\right)$$

is an orthogonal expansion for every $f \in PW_B$.

(Simplifications effected by special choices for k were noted by Grechikhin (1969).)

13.8 Find a band-pass "function plus derivative" sampling series (the reconstruction functions for this and several other multi-band sampling series have been calculated by Kempski (1995) using symbolic computation).

13.9 There is a striking similarity between the band-pass function plus Hilbert transform series (13.6.11) and the low-pass series (12.2.8). Show that the band-pass version reduces to (12.2.8) (as it should) when the gap between the band components closes, i.e. when $w_0 \to 0$.

13.10 Discuss the possibility of taking $M_j(x) := x^{j-1}$, $(j = 1, \dots, N)$ in Theorem 13.9.

13.11 Let the transformations S_j be as in Theorem 13.9. Formulate a definition of stable sampling with respect to these transformations, and make sure that it applies to the series of Theorems 13.9 and 13.13.

14

MULTI-DIMENSIONAL SAMPLING

Our previous work has largely been concerned with sampling theories in one dimension. Now we are going to extend matters to higher dimensions.

We wish to find sampling representations for functions $f : \mathbb{R}^N \mapsto \mathbb{C}$ that are band-limited to a region $B \subset \mathbb{R}^N$. Let us recall that an appropriate class of functions is the multi-dimensional Paley–Wiener space PW_B, already introduced in §6.6 and defined to be $\{f : f \in L^2(\mathbb{R}^N) \cap C(\mathbb{R}^N); \operatorname{supp} f^\wedge \subseteq \overline{B}\}$. In analogy with the one-dimensional case, it is a Hilbert space with reproducing kernel, and is isometrically isomorphic to $L^2(B)$.

The idea of stable sampling can be formulated in direct analogy to the one-dimensional case. We say that $\{\lambda_n\} \subset \mathbb{R}^N$, $n \in \mathbb{Z}^N$, is a set of stable sampling for PW_B if there is a constant K such that, for every $f \in PW_B$,

$$\|f\|^2_{L^2(\mathbb{R}^N)} \leq K \sum_{n \in \mathbb{Z}^N} |f(\lambda_n)|^2.$$

Let us also recall that the Nyquist–Landau (N–L for short) sample point density is defined to be $m(B)/(2\pi)^N$. In a multi-dimensional form of Theorem 17.9, Landau (1967b) proves that stable sampling is not possible at densities below this N–L value. An important consideration here, as for one-dimensional sampling, is whether this minimal sampling density can be achieved. Sampling at a larger density is called over-sampling as before.

Naturally, a hyper-rectangle can always be found to enclose a given bounded band-region $B \subset \mathbb{R}^N$. In that case a function band-limited to B can be reconstructed by the simple sampling theorem of §14.2 below, but usually at the price of over-sampling. The nature of B will play a vital rôle if the N–L rate is to be achieved.

In Theorem 14.4, the main result of the chapter, we shall find that virtually all of the nice features of the one-dimensional regular case can be retained, provided that the band-region B has the special property that it, and copies of it obtained from a group of translations, form a partition of $\mathbb{R}^N$ (this is a multi-dimensional version of *tessellation* introduced in Chapter 13). We shall only treat cases in which B is bounded, and has finitely many disjoint components, each of which is of positive N-dimensional Lebesgue measure. One could relax these conditions somewhat, but at the expense of introducing certain pathological cases that are not of direct interest to us.

Indeed, Theorem 14.4 allows regularity of the sample point distribution, it allows the N–L minimal sampling density to be achieved and, furthermore, the sampling series is an expansion in an orthonormal basis for PW_B. The theorem also contains a sharp aliasing error bound and an example of an extremal in analogy with the result for ordinary sampling in Theorem 11.7.

The regular hexagon is an example of a suitable B (as in Example 14.5). However, there is no reason why B should be simply connected. The reader will have no difficulty in constructing a more exotic example by taking a square, removing a disc, say, from its interior and placing the disc (prudently) elsewhere; and so providing a suitable B having two disjoint connected components, one simply connected and the other not.

Of course there are simple examples of band-regions B for which Theorem 14.4 does not apply. The classical example is the disc in $\mathbb{R}^2$, or more generally the ball in $\mathbb{R}^N$. A sampling theorem can be obtained when B is a disc, but it does not achieve the N–L sample point density; even here though, there is an algorithmic method for improving matters (references can be found in Example 14.5 below).

14.1   Remarks on multi-dimensional Fourier analysis

A small amount of multi-dimensional Fourier analysis will be needed in this chapter, which we shall take for granted; it consists of straightforward extensions to higher dimensions of facts already reviewed in Chapter 2. An excellent account of Fourier analysis on $\mathbb{R}^N$ can be found in Stein and Weiss (1971). Here, just a few items of appropriate notation will be introduced.

Let t and x continue to denote time and frequency variables, but now let us take them to be the vectors $t = (t_1, \ldots, t_N)$ and $x = (x_1, \ldots, x_N)$ respectively. As usual, $t \cdot x = t_1 x_1 + \cdots + t_N x_N$ denotes the ordinary scalar product. Let $m(S)$ denote the N-dimensional Lebesgue measure of the set $S \subset \mathbb{R}^N$.

Let $w = (\pi w_1, \ldots, \pi w_N)$, in which $w_j > 0$, $j = 1, \ldots, N$, and for $n = (n_1, \ldots, n_N) \in \mathbb{Z}^N$ let $n/w := (n_1/w_1, \ldots, n_N/w_N)$. Also,

$$R_w := [-\pi w_1, \pi w_1) \times \cdots \times [-\pi w_N, \pi w_N)$$

will be called a *hyper-rectangle*. It follows by a standard extension of the one-dimensional result that the set $\{e^{ix \cdot n/w} / \sqrt{m(R_w)}\}$, $n \in \mathbb{Z}^N$, is an orthonormal basis for $L^2(R_w)$.

If f is multiply periodic with the periods πw_j in the variable x_j, $j = 1, \ldots, N$, we can associate with it a formal Fourier series

$$f(x) \sim \frac{1}{m(R_w)} \sum_{n \in \mathbb{Z}^N} C_n e^{ix \cdot n/w},$$

where $C_n = \int_{R_w} f(x) e^{-ix \cdot n/w}\, dx$.

When the pointwise convergence of a multi-sum is required it will be understood in the sense of the limit of rectangular partial sums, i.e.

$$\lim_{Q_1 \to \infty} \sum_{|n_1| \leq Q_1} \cdots \lim_{Q_N \to \infty} \sum_{|n_N| \leq Q_N}.$$

The *Fourier transform* of a function f is defined formally by

$$f^\wedge(x) = (\mathcal{F}f)(x) = \frac{1}{(2\pi)^{N/2}} \int_\mathbb{R} f(t) e^{-ix\cdot t}\, dt,$$

and the formal *Fourier inversion formula* is

$$f(t) = (\mathcal{F}^{-1}f^\wedge)(t) = \frac{1}{(2\pi)^{N/2}} \int_\mathbb{R} f^\wedge(x) e^{ix\cdot t}\, dx.$$

14.2   The rectangular case

The simplest case is when B is a hyper-rectangle R_w. Then we have the following N-dimensional sampling theorem, a standard extension of the one-dimensional case. It can be proved directly using the orthonormal set of the previous section, but since it is a special case of the more general Theorem 14.4 of the next section we shall not give a separate proof.

Theorem 14.1 *Let f belong to the Paley–Wiener space PW_{R_w}. Then*

$$f(t) = \sum_{n\in\mathbb{Z}^N} f\left(\frac{n}{w}\right) \text{sinc}(w_1 t_1 - n_1) \ldots \text{sinc}(w_N t_N - n_N),$$

convergence being in the norm of $L^2(\mathbb{R}^N)$, and globally uniform over $\mathbb{R}^N$.

Note that the sampling density is $w_1 \ldots w_N$, which is the Nyquist–Landau minimal sampling density $m(B)/(2\pi)^N$.

14.3   Regular multi-dimensional sampling

Let $v_1,\ldots,v_N$ be vectors which span $\mathbb{R}^N$, and let V be the $N \times N$ matrix whose jth column is v_j. Let $u_1,\ldots,u_N$ be the associated set of biorthogonal vectors such that

$$u_i \cdot v_j = 2\pi\delta_{ij}, \qquad i,j = 1,\ldots,N. \tag{14.3.1}$$

Let U be the matrix whose ith row is u_i, so that $U = 2\pi V^{-1}$.
Let $s = (s_1,\ldots,s_N) \in \mathbb{Z}^N$. The "sampling lattice" is the set

$$\mathbb{L} := \{l: l = s_1 v_1 + \cdots + s_N v_N = Vs;\ s \in \mathbb{Z}^N\}. \tag{14.3.2}$$

Of course $\mathbb{L}$ is a discrete subgroup of the additive group $\mathbb{R}^N$; it is isomorphic to the *free Abelian group on N generators*. A set of sampling points with this property will be called *regular*.
Similarly, we define

$$\Xi := \{\xi: \xi = q_1 u_1 + \cdots + q_N u_N = U^T q;\ q \in \mathbb{Z}^N\}, \tag{14.3.3}$$

which is also a subgroup of $\mathbb{R}^N$, and is also isomorphic to the free Abelian group on N generators. This group is called the *annihilator*[1] of $\mathbb{L}$ and is sometimes written $\mathbb{L}^\perp$.

We now form the quotient group $\mathbb{R}^N/\Xi$ consisting of cosets $\{\bar{t} = t + \Xi : t \in \mathbb{R}^N\}$ (in which, as usual, $t + \Xi$ means $\{t + \xi : \xi \in \Xi\}$). We refer to $\bar{t}$ as the coset containing t; any member of the coset $\bar{t}$ determines it, and is therefore called a *representative of* $\bar{t}$.

A set B is called a *complete set of coset representatives* of $\mathbb{R}^N/\Xi$ if its membership consists of exactly one representative of every coset. In the theory of translation groups such a set B is called a *fundamental region* for Ξ.

The following three facts follow from the definitions we have just made, and are left as exercises for the reader.

1. A typical fundamental region for Ξ is the hyper-parallelepiped P so placed that one corner is at the origin, and the N edges which meet at this corner are the vectors $u_1, \ldots, u_N$ with their tails placed at the origin. Also we must choose P to be "half-closed, half-open", and this can be arranged by including, for example, those N faces of P which meet at the origin, and excluding from P the remaining N, which do not.

2. Every fundamental region B tessellates $\mathbb{R}^N$ when translated by all members of Ξ.

3. The exponential functions $\{e^{ix \cdot l}\}$, $l \in \mathbb{L}$, are periodic functions of $x \in \mathbb{R}^N$, with periods $u_1, \ldots, u_N$. It follows that

$$\int_B e^{ix \cdot l} \, dx = \int_D e^{ix \cdot l} \, dx, \qquad l \in \mathbb{L},$$

for any two fundamental regions B and D.

Lemma 14.2 *If B is any fundamental region for Ξ, then $m(B) = \det U$.*

Proof   A standard formula of geometry gives $m(P) = \det U$. Further,

$$m(P) = \int_{\mathbb{R}^N} \chi_P(x) \, dx = \sum_{\xi \in \Xi} \int_B \chi_P(x + \xi) \, dx$$

$$= \int_B \sum_{\xi \in \Xi} \chi_P(x + \xi) \, dx = m(B),$$

there being finitely many terms only in these sums.[2]

Next we come to a lemma which is of basic importance for multi-dimensional sampling.

---

[1] So called because $e^{i\xi \cdot l} \equiv 1$; this means that the set of functions $\{e^{ix \cdot l}\}$, which will assume an importnat rôle in Lemma 14.3, has its dependence on $\mathbb{L}$ annihilated by evaluating any member on Ξ.

[2] Readers wishing to consider the case where B has infinitely many components will be able to justify the interchange of sum and integral using the monotone convergence theorem.

Lemma 14.3 *Let B be any fundamental region for Ξ. Let us define g, a reconstruction function, by*

$$g(t) := \frac{1}{m(B)} \int_B e^{ix \cdot t}\, dx.$$

Then:

1. $g(l) = \begin{cases} 0, & l \in \mathbb{L},\ l \neq 0 \\ 1, & l = 0. \end{cases}$

2. *The set $\{e^{ix \cdot l}/\sqrt{m(B)}\}$, $l \in \mathbb{L}$, is an orthonormal basis for $L^2(B)$.*

3. *The inverse fourier transforms of the complex conjugates of the functions in part 2 (when given the value 0 outside B) are $\{\sqrt{m(B)/(2\pi)^N}\ g(t - l)\}$, $l \in \mathbb{L}$.*

Proof   Part 3 is straightforward.

As to part 1, first we clearly have $g(0) = 1$. Now if B is a fundamental region for Ξ, then clearly so is B translated by any $y \in \mathbb{R}^N$ (we denote such a translate by $B - y$). Hence, for any $l \in \mathbb{L}$,

$$\int_B e^{ix \cdot l}\, dx = \int_{B-y} e^{ix \cdot l}\, dx = \int_B e^{i(x+y) \cdot l}\, dx = e^{iy \cdot l} \int_B e^{ix \cdot l}\, dx.$$

Therefore either $\int_B e^{ix \cdot l}\, dx = 0$, or $e^{iy \cdot l} = 1$. The latter case can only hold for every y if $l = 0$. This proves part 1, and since the orthogonality in part 2 involves an integral of the form $\int_B e^{ix \cdot (l-l')}\, dx$, the same method establishes this part of the proof.

It remains to show that $\{e^{ix \cdot l}\}$, indexed by $l \in \mathbb{L}$, is complete in $L^2(B)$. Such a result is certainly plausible on intuitive grounds. We know that completeness holds in each dimension separately and this strongly suggests the result, certainly for $L^2(P)$, and general ideas of periodicity suggest the general result. Unfortunately, there seems to be no elementary proof that does not tend to obscure the basic ideas with cumbersome detail. However, there are clean and very general proofs at a higher conceptual level that we can reach here; for example, in Loomis (1953, p. 154) and in the contribution by Dodson and Beaty to the book H–S.

Theorem 14.4 *Let $f \in F = \{f : f \in L^2(\mathbb{R}^N) \cap C(\mathbb{R}^N); f^\wedge \in L(\mathbb{R}^N)\}$. Let $\mathbb{L}$ and Ξ be as in (14.3.2) and (14.3.3), let g be as in Lemma 14.3 and let B be any fundamental region for Ξ. Then:*

1. *The aliasing error A satisfies*

$$A(t) = \left| f(t) - \sum_{l \in \mathbb{L}} f(l) g(t - l) \right| \leq \frac{2}{(2\pi)^{N/2}} \int_{\mathbb{R}^N \setminus B} |f^\wedge(x)|\, dx. \quad (14.3.4)$$

2. *The constant $2/(2\pi)^{N/2}$ in the error bound is best possible.*
3. *If $f \in PW_B$, the N-dimensional regular sampling representation*

$$f(t) = \sum_{l \in \mathbb{L}} f(l) g(t - l)$$

holds, with convergence in the norm of $L^2(\mathbb{R}^N)$, and pointwise and globally uniformly over $\mathbb{R}^N$.

The set $\{g(t - l)\}$, $l \in \mathbb{L}$, is an orthonormal basis for PW_B. The associated Parseval relation is

$$\|f\|^2 = \frac{(2\pi)^N}{m(B)} \sum_{l \in \mathbb{L}} |f(l)|^2.$$

Proof The ideas of the proof owe much to the methods surrounding Poisson's summation formula. To prove part 1, we observe first that for any f satisfying our hypotheses, $\sum_{\xi \in \Xi} f^\wedge(x + \xi)$ converges absolutely in the norm of $L(B)$ to a function H, say (which will clearly be periodic with periods $u_1, \ldots, u_N$). To see this, we have

$$\sum_{\xi \in \Xi} \int_B |f^\wedge(x + \xi)| \, dx = \sum_{\xi \in \Xi} \int_{B - \xi} |f^\wedge(x)| \, dx = \int_{\mathbb{R}^N} |f^\wedge(x)| \, dx = \|f^\wedge\|_1,$$

so by the standard theorem on interchange of sum and integral we have

$$\int_B \sum_{\xi \in \Xi} |f^\wedge(x + \xi)| \, dx < \infty,$$

which proves the assertion. Now the formal Fourier series representation for H is

$$H(x) = \sum_{\xi \in \Xi} f^\wedge(x + \xi) \sim \frac{1}{\sqrt{m(B)}} \sum_{l \in \mathbb{L}} c_l e^{-ix \cdot l} \tag{14.3.5}$$

with coefficients

$$c_l = \frac{1}{\sqrt{m(B)}} \int_B \sum_{\xi \in \Xi} f^\wedge(x + \xi) e^{ix \cdot l} \, dx.$$

On interchanging sum and integral as before, changing variable and using the biorthogonality condition (14.3.1) we find that c_l is given by

$$\frac{1}{\sqrt{m(B)}} \sum_{\xi \in \Xi} \int_{B - \xi} f^\wedge(x) e^{i(x - \xi) \cdot l} \, dx = \frac{1}{\sqrt{m(B)}} \sum_{\xi \in \Xi} \int_{B - \xi} f^\wedge(x) e^{ix \cdot l} \, dx$$

$$= \frac{1}{\sqrt{m(B)}} \int_{\mathbb{R}^N} f^\wedge(x) e^{ix \cdot l} \, dx = \frac{(2\pi)^{N/2}}{\sqrt{m(B)}} f(l).$$

After substituting for c_l we multiply both sides of (14.3.5) by $(2\pi)^{-N/2} e^{ix \cdot t}$ and integrate term by term over B. The right-hand side gives

$$\sum_{l\in L} f(l)\left\{\frac{1}{m(B)}\int_B e^{ix\cdot(t-l)}\,dx\right\} = \sum_{l\in L} f(l)g(t-l).$$

The left-hand side gives

$$(2\pi)^{-N/2}\int_B \sum_{\xi\in\Xi} f^\wedge(x+\xi)e^{ix\cdot t}\,dx = \sum_{\xi\in\Xi} e^{-i\xi\cdot t}(2\pi)^{-N/2}\int_{B-\xi} f^\wedge(x)e^{ix\cdot t}\,dx.$$

Now to this let us add and subtract

$$f(t) = (2\pi)^{-N/2}\sum_{\xi\in\Xi}\int_{B-\xi} f^\wedge(x)e^{ix\cdot t}\,dx,$$

so that our left-hand side becomes

$$f(t) + (2\pi)^{-N/2}\sum_{\xi\in\Xi}\left(e^{-i\xi\cdot t}-1\right)\int_{B-\xi} f^\wedge(x)e^{ix\cdot t}\,dx.$$

This summand vanishes when $\xi = 0$, and when $\xi \neq 0$ we have $|e^{-i\xi\cdot t}-1| \leq 2$. Our left- and right-hand sides can now be combined to assemble $A(t)$ together with an error term, and then a simple estimate gives

$$|A(t)| \leq \frac{2}{(2\pi)^{N/2}}\sum_{\substack{\xi\in\Xi\\ \xi\neq 0}}\int_{B-\xi} |f^\wedge(x)|\,dx = \frac{2}{(2\pi)^{N/2}}\int_{\mathbb{R}^N\setminus B} |f^\wedge(x)|\,dx.$$

This completes the proof of part 1.

To prove part 2 we must find an extremal for the error bound; that is, we must find a member X of F such that, given any fundamental region B for Ξ, inequality (14.3.4) applied to X reduces to an equality for at least one value of t. We shall find our extremal in a Paley–Wiener space with associated band-region B', obtained by augmenting B as follows. Let

$$\Xi' = \{\xi' : \xi' = 2q_1 u_1 + q_2 u_2 + \cdots + q_N u_N;\ q \in \mathbb{Z}^N\}$$

and let B' be any fundamental region for Ξ' (here we have chosen to double the length of u_1, but we could have chosen any one of $u_1,\ldots,u_N$ for this purpose; this, together with the wide choice available for B', indicates that our construction of an extremal is by no means unique).

It is clear that $m(B') = 2m(B)$. The following definitions are completely analogous to those we have already made in connection with B.

Put

$$u'_1 = 2u_1,\ u'_2 = u_2,\ \ldots,\ u'_N = u_N;$$
$$v'_1 = \frac{1}{2}v_1,\ v'_2 = v_2,\ \ldots,\ v'_N = v_N.$$

Then we have the biorthogonality relation

$$u'_i \cdot v'_j = 2\pi \delta_{ij}, \qquad i, j = 1, \ldots, N.$$

Put

$$\mathbb{L}' = \{l' : \ l' = s_1 v'_1 + \cdots + s_N v'_N; \ s \in \mathbb{Z}^N\}.$$

Then, as in Lemma 14.3, the set

$$\left\{ e^{ix \cdot l'} / \sqrt{m(B')} \right\}, \qquad l' \in \mathbb{L}' \tag{14.3.6}$$

is an orthonormal basis for $L^2(B')$. We now show that an extremal for the error bound (14.3.4) is

$$X(t) = \chi_{B'}^{\vee} \left(t - (v'_1 + \cdots + v'_N) \right). \tag{14.3.7}$$

Now

$$X(v'_1 + \cdots + v'_N) = \frac{1}{(2\pi)^{N/2}} \int_{B'} dx = \frac{m(B')}{(2\pi)^{N/2}} = \frac{2}{(2\pi)^{N/2}} \, m(B),$$

and

$$X(l) = \chi_{B'}^{\vee} \left(l - (v'_1 + \cdots + v'_N) \right)$$
$$= \frac{1}{(2\pi)^{N/2}} \int_{B'} e^{ix \cdot [(2s_1 - 1)v'_1 + (s_2 - 1)v'_2 + \cdots + (s_N - 1)v'_n]} \, dx.$$

Now this integrand is one of the exponentials in the basis (14.3.6), but not the one that is identically 1 over B'; hence, arguing as in Lemma 14.3 part 1, $X(l) = 0$, $l \in \mathbb{L}$. This shows that the aliasing error $A(t)$ for X reduces to just $|X(t)|$.

Now the right-hand side of (14.3.4) is

$$\frac{2}{(2\pi)^{N/2}} \int_{\mathbb{R}^N \backslash B} \left| \{\chi_{B'}^{\vee} \left(t - (v'_1 + \cdots + v'_N) \right)\}^{\wedge} (x) \right| dx$$

$$= \frac{2}{(2\pi)^{N/2}} \int_{\mathbb{R}^N \backslash B} \chi_{B'}(x) \left| e^{-ix \cdot [v'_1 + \cdots + v'_N]} \right| dx = \frac{2}{(2\pi)^{N/2}} \int_{(\mathbb{R}^N \backslash B) \cap B'} dx$$

$$= \frac{2}{(2\pi)^{N/2}} \int_{B' \backslash B} dx = \frac{2}{(2\pi)^{N/2}} m(B),$$

and so equality in (14.3.4) is achieved when $t = v'_1 + \cdots + v'_N$.

To prove part 3 we could observe first that if $f \in PW_B$ the error bound $A(t)$ vanishes, and this gives the pointwise convergence of the series. But the more comprehensive result is obtained by noting that the function set of Lemma 14.3, part 3, is an orthonormal basis for PW_B (this follows from the usual Fourier duality), and that it is easily verified that the coefficients for f in this set are

$\{\sqrt{(2\pi)^N/m(B)}\,f(l)\}$. This shows that the series converges in norm, and gives the Parseval relation. The remaining convergence properties follow from the Convergence Principle of §6.6.

We noted at the beginning of this chapter that the idea of stability generalises readily to higher dimensions. To our present theorem we can add the obsevation, therefore, that the Parseval relation of part 3 shows the sampling lattice $\mathbb{L}$ to be a set of stable sampling for PW_B.

It is natural to ask whether there are sampling series for functions belonging to spaces PW_B for which B does not satisfy the requirements of Theorem 14.4; that is, if it is not a fundamental region for a translation group. There is no comprehensive theory, but progress can be made in certain directions. Naturally one would hope to retain as many nice features of the regular case as possible, but it is inevitable that something will be missing.

The example to follow, that of *hexagonal sampling* for functions with a circular band-region, illustrates one of the possibilities. Other examples can be found in the problem set, particularly in Problem 14.3. These examples are interesting because of the methods they employ, and because attempts to generalize them will suggest directions for further study to the interested reader.

Example 14.5 (Hexagonal Sampling) We wish to sample and reconstruct a function defined on $\mathbb{R}^2$ whose band-region is the circular disk D (of unit radius for simplicity). Since D is not a tessellating figure for the plane, the best we can do is to choose that tessellating figure of smallest area which contains D, a regular hexagon H say, and endow it with the rôle of assumed band-region. It is easy to calculate that the centres of the tessellating hexagons comprise the set

$$\Xi = \{n_1u_1 + n_2u_2\}, \qquad n = (n_1, n_2) \in \mathbb{Z}^2,$$

where $u_1 = \sqrt{3}e_1 - e_2$ and $u_2 = 2e_2$ and $\{e_1, e_2\}$ is the usual orthogonal unit vector basis for $\mathbb{R}^2$. We also have $v_1 = (2\pi/\sqrt{3})e_1$ and $v_2 = (\pi/\sqrt{3}e_1 + \pi e_2)$.

Now Theorem 14.4 can be applied using H for the band-region B. The reconstruction function g is calculated as in Lemma 14.3; the result is

$$\frac{2t_1 \cos(t_1/\sqrt{3}) \cos t_2 - 2t_1 \cos(2t_1/\sqrt{3}) - 2\sqrt{3}\sin(t_1/\sqrt{3}) \sin t_2}{t_1(t_1^2 - 3t_2^2)}.$$

Since the N–L sampling rate is not achieved in this example, there remains a certain redundancy in the sample points, and an obvious question is whether there is some other sampling and reconstruction process which can remove some of this redundancy. An algorithmic method has indeed been given for doing this (Cheung 1993, Ch. 3, §3.6).

PROBLEMS

14.1 A function $f : \mathbb{R}^N \mapsto \mathbb{R}$ is said to have *radial symmetry*, or simply to be *radial*, if $f(t_1) = f(t_2)$ whenever $|t_1| = |t_2|$. Show that if f is radial, so is $f^\wedge$.

Show that if $r = |t|$, $\rho = |x|$ and $f(t) = F(r)$, $f^\wedge(x) = F^\wedge(\rho)$, then

$$\rho^{\frac{n}{2}-1} F^\wedge(\rho) = \int_0^\infty r[r^{\frac{n}{2}-1}F(r)] J_{\frac{n}{2}-1}(\rho r)\, dr,$$

so that $\rho^{(n-1)/2} F^\wedge(\rho)$ is the Hankel transform of $r^{(n-1)/2} f(r)$ of order $n/2 - 1$.

Formulate a sampling theorem for band-limited radial functions.

14.2 Find an aliasing error bound if the actual band-region is a circle, and if the assumed band-region is taken to be a square of the same area.

14.3 Consider a square of side length three units centred at the origin, its sides parallel to the co-ordinate axes. Let this square be subdivided into nine smaller equal squares. Pick out some of these smaller squares so that their union forms the band-region B for a two-dimensional Paley–Wiener space, and develop a appropriate sampling and reconstruction series.

One possible case is when B consists of the middle square of the top and bottom row, and the two end squares of the middle row. Show that a complete orthonormal basis for $L^2(B)$ is

$$\{e^{2ix\cdot n}\} \cup \{e^{2ix\cdot n}\, e^{ix_1/2}\} \cup \{e^{2ix\cdot n} e^{ix_2/2}\} \cup \{e^{2ix\cdot n} e^{i(x_1+x_2)/2}\},$$

where $n = (n_1, n_2) \in \mathbb{Z}^2$ (each member is normalized by the factor $1/2\pi$), and the reconstruction function is

$$g(t) = \cos\frac{\pi}{2}(t_1 + t_2)\cos\frac{\pi}{2}(t_1 - t_2)\,\mathrm{sinc}\,\frac{t_1}{2}\,\mathrm{sinc}\,\frac{t_2}{2}.$$

14.4 Show that when $\mathcal{A}$ is the hyperparallelepiped P of §14.3, (14.3.7) gives

$$X(t) = \frac{\sin 2\pi(u_1 \cdot t - \frac{1}{2})}{2\pi\left(u_1 \cdot t - \frac{1}{2}\right)} \prod_{j=2}^N \frac{\sin \pi u_j \cdot t}{\pi u_j \cdot t}.$$

It is interesting to note that this extremal is not just a product of one-dimensional extremals. Show that, in fact, such a product is not extremal.

14.5 Show by using an appropriate basis for $L^2([0,\pi] \times [0,\pi])$ that if

$$f(s,t) = \int_0^\pi \int_0^\pi g(x,y) \cos xt \cos ys\, dx\, dy \tag{14.3.8}$$

for some $g \in L^2([0,\pi] \times [0,\pi])$, then there is the following two-dimensional counterpart of the "shifted" cardinal series:

$$f(s,t) = \sum_{(m,n)\in\mathbb{Z}^2} f(m+\tfrac{1}{2}, n+\tfrac{1}{2}) \operatorname{sinc}(s-m-\tfrac{1}{2}) \operatorname{sinc}(t-n-\tfrac{1}{2}).$$

14.6 Let Δ denote the region bounded by the lines $x = 0$, $y = x$ and $y = \pi$, and let χ_Δ denote its characteristic function. In (14.3.8) take $g(x,y) = J_0(b\sqrt{y^2 - x^2})\chi_\Delta(x,y)$. Using Appendix A, no. 19, show that

$$f(s,t) = \Omega(s,t) + \Omega(s,-t),$$

where

$$\Omega(X,Y) = \frac{\sin^2\left(\frac{\pi}{2}\left(\sqrt{b^2+X^2}+Y\right)\right)}{\sqrt{b^2+X^2}\left(\sqrt{b^2+X^2}+Y\right)},$$

and consequently

$$\Omega(s,t) + \Omega(s,-t) = \sum_{(m,n)\in\mathbb{Z}^2} \{\Omega(m+\tfrac{1}{2}, n+\tfrac{1}{2}) + \Omega(m+\tfrac{1}{2}, -n-\tfrac{1}{2})\}$$

$$\times \operatorname{sinc}(s-m-\tfrac{1}{2})\operatorname{sinc}(t-m-\tfrac{1}{2}).$$

Obtain the special case

$$\frac{\pi^2}{2}\operatorname{sinc}^2\frac{\pi s}{2} = \sum_{(m,n)\in\mathbb{Z}^2} \frac{(-1)^n}{\pi(m+\tfrac{1}{2})(n+\tfrac{1}{2})}$$

$$\times \left\{\frac{\cos^2\left(\frac{\pi}{2}(m+n)\right)}{m+n+1} + \frac{\sin^2\left(\frac{\pi}{2}(m-n)\right)}{m-n}\right\}\operatorname{sinc}(s-m-\tfrac{1}{2}).$$

Two further special cases are the double series representation

$$\pi^4 = 2\sum_{(m,n)\in\mathbb{Z}^2} \frac{(-1)^{m+n}}{(m+\tfrac{1}{2})^2(n+\tfrac{1}{2})}\left\{\frac{\cos^2\left(\frac{\pi}{2}(m+n)\right)}{m+n+1} + \frac{\sin^2\left(\frac{\pi}{2}(m-n)\right)}{m-n}\right\}$$

and the well known series

$$\pi = 8\sum_{n=0}^{\infty} \frac{1}{(4n+1)(4n+3)}$$

(Zayed 1993b, p. 222).

15

SAMPLING AND EIGENVALUE PROBLEMS

Suppose that $\mathcal{L}$ denotes a differential operator on an L^2 space, and that an appropriate solution $Y(x, \lambda)$ of the eigenvalue equation $\mathcal{L}y = \lambda y$ can be chosen so that $\{Y(x, \lambda_n)\}$ are the eigenfunctions corresponding to eigenvalues $\{\lambda_n\}$ (the relevant definitions can be found in §15.1). If there is an expansion theory associated with these eigenfunctions, then under suitable conditions there is a companion expansion which has the nature of a sampling series. This is the direction in which Kramer suggested that his Lemma (Chapter 8) should be developed. The aim of the present chapter is to introduce some of the many known ways in which this procedure can be made precise.

The cardinal series arises very simply in this way, from the eigenvalue problem

$$-iy' = \lambda y$$
$$y(-\pi) = y(\pi).$$

We find that $e^{ix\lambda}$ is a solution of the differential equation, and satisfies the boundary condition if $\lambda_n = n$, $n \in \mathbb{Z}$. The corresponding eigenfunctions are $\{e^{ixn}\}$, and one now proceeds to the cardinal series for Paley–Wiener functions, as described at the beginning of Chapter 8.

In Corollary 5.3 we saw that the cardinal series is a limiting case of the Lagrange interpolation formula, and in §10.2 we met Lagrange interpolation again in the context of irregular sampling. It is natural to ask if a sampling series associated with an eigenvalue problem can also be viewed as a form of Lagrange interpolation. In Theorem 15.6 and Examples 15.11 and 15.13 we shall find that, under suitable conditions, there is indeed such a connection.

Throughout this chapter the reader will be able to interpret various norms and inner products in context without difficulty.

15.1 Preliminary facts

We shall need the following facts and definitions concerning operators and their spectra. A very readable introduction to this topic can be found in the book of Lorch (1962). Let $\mathcal{T}$ denote an operator on a Hilbert space H and let $\mathfrak{D}(\mathcal{T})$, the domain of $\mathcal{T}$, be dense in H. Then we can associate with $\mathcal{T}$ an *adjoint transformation* $\mathcal{T}^*$ with domain $\mathfrak{D}(\mathcal{T}^*) \subseteq$ H and satisfying $\langle \mathcal{T}u, v \rangle = \langle u, \mathcal{T}^*v \rangle$, where $u \in \mathfrak{D}(\mathcal{T})$ and $v \in \mathfrak{D}(\mathcal{T}^*)$. Of course $\mathfrak{D}(\mathcal{T}^*)$ will not normally coincide with $\mathfrak{D}(\mathcal{T})$. However, it may happen that $\mathcal{T}$ and $\mathcal{T}^*$ have the same domain and are identical; in that case $\mathcal{T}$ is called *self-adjoint*. Thus, the condition for self-adjointness is

$$\langle Tu, v \rangle = \langle u, Tv \rangle \quad \text{whenever} \quad u, v \in \mathfrak{D}(T).$$

It will be instructive for the reader to check the facts in the following:

Example 15.1 If $\mathcal{D}$ denotes differentiation, the operator $T := -\mathcal{D}^2$ is:
(a) self-adjoint on $L^2(0, \pi)$ if $\mathfrak{D}(T) = \{y \colon y \in C^2[0, \pi]; y'(0) = y'(\pi) = 0\}$;
(b) not self-adjoint on $L^2(-\pi, \pi)$ if $\mathfrak{D}(T) = \{y \colon y \in C^2[-\pi, \pi]; y'(-\pi) = 0, y(-\pi) + y'(\pi) = 0\}$.

This example reminds us that $\mathfrak{D}(T)$ has a decisive influence on the nature of T.

Because we are interested mostly in the self-adjoint case the information contained in Theorem 15.3 below will be of basic importance. First we need some definitions.

Definition 15.2 *Let T be an operator defined on a domain that is dense in a Hilbert space H. A scalar λ belongs to the point spectrum of T if there exists a non-null element f of $\mathfrak{D}(T)$ such that $(T - \lambda \mathcal{I})f = 0$, where $\mathcal{I}$ is the identity operator. The operator T is said to have pure point spectrum if the collection of all those f's which satisfy $(T - \lambda \mathcal{I})f = 0$ for some λ (depending on f) spans the space H. In the case of pure point spectrum, any f for which $Tf = \lambda \mathcal{I}f$ for a particular λ is called an eigenfunction belonging to the eigenvalue λ. The collection of all eigenfunctions belonging to a particular λ spans a subspace N_λ of H called the eigenspace belonging to λ; the multiplicity of an eigenvalue λ is defined to be the dimension of N_λ. When the dimension is 1, λ is said to be a simple eigenvalue.*

The information contained in the following theorem can be found in Lorch (1962, Ch. V).

Theorem 15.3 *Let T be self-adjoint and be of the pure point spectrum type. Then all its eigenvalues are real, and eigenfunctions corresponding to distinct eigenvalues are orthogonal.*

We now introduce the Nth order formal linear differential operator $\mathcal{L}$, given by

$$\mathcal{L}y := p_0 y^{(N)} + p_1 y^{(N-1)} + \cdots + p_N y,$$

where $p_j(x) \in C^{N-j}[a, b]$, $j = 0, \ldots, N$, and $p_0 \neq 0$ on $[a, b]$. Also we have the *boundary forms*

$$\mathcal{B}y := \mathcal{B}_j y = \alpha_{j1} y(a) + \cdots + \alpha_{jN} y^{(N-1)}(a)$$
$$+ \beta_{j1} y(b) + \cdots + \beta_{jN} y^{(N-1)}(b), \qquad j = 1, \ldots, N. \quad (15.1.1)$$

The operators to be considered will be of the type $T := \mathcal{L}$ with $\mathfrak{D}(T) = \{f \colon f \in C^N[a, b]; \mathcal{B}y = 0\}$. As such, T is densely defined on $L^2(a, b)$. An *eigenvalue problem* consists in determining the eigenvalues and eigenfunctions of an operator T of this type. If T is self-adjoint we refer to a *self-adjoint eigenvalue problem*. We shall write our eigenvalue problems in the form

$$\mathcal{L}y = \lambda y,$$
$$\mathcal{B}_j y = 0, \qquad j = 1, \ldots, N,$$

and refer to the two parts as the *eigenvalue equation* and the *boundary conditions* respectively.

The theory of differential equations guarantees that there exist N linearly independent solutions, $Y_j(x, \lambda)$, $j = 1, \ldots, N$, say, of the eigenvalue equation $\mathcal{L}y = \lambda y$ which constitute a *fundamental system* of solutions for each $\lambda \in \mathbb{C}$. The N-dimensional vector space spanned by a fundamental system is called the *solution space* of the equation.

While a self-adjoint operator is not, in general, necessarily of the pure point spectrum type, it can be proved (as in Coddington and Levinson (1955 p. 189) for example) that if the operator is defined by an eigenvalue problem as described above, and is self-adjoint on the domain specified by the boundary conditions, then it is of the pure point spectrum type and of course Theorem 15.3 applies.

A final preliminary fact of a different nature that will be needed is the following simple special case of Hadamard's Factorization Theorem (the complete theorem can be found in Young (1980, p. 65)).

Theorem 15.4 *If $f(\lambda)$ is an entire function of order $\frac{1}{2}$, and has simple zeros at $\{\lambda_n\}$, then*

$$f(\lambda) = CL(\lambda),$$

where

$$L(\lambda) = \prod_{k=0}^{\infty} \left(1 - \frac{\lambda}{\lambda_k}\right) \tag{15.1.2}$$

if f does not vanish at the origin, and

$$L(\lambda) = \lambda \prod_{k=1}^{\infty} \left(1 - \frac{\lambda}{\lambda_k}\right) \tag{15.1.3}$$

if f has a simple zero at the origin, and C is a constant depending on f.

15.2 Direct and inverse Sturm–Liouville problems

Suppose that the differential operator $\mathcal{L}$ described in the previous section is of the second order, and is of the form $\mathcal{L}y := -y'' + q(x)y$, where q is real valued and continuous on a bounded interval $[a, b]$. Also, let the boundary conditions be such that $\mathcal{B}_1$ is a condition at one end-point only of $[a, b]$, and $\mathcal{B}_2$ is a condition at the other. These are called *separated boundary conditions*.

This kind of eigenvalue problem is called a *regular Sturm–Liouville eigenvalue problem*, and can be written

$$\mathcal{L}y := -y'' + q(x)y = \lambda y \tag{15.2.1}$$
$$\cos \alpha y(a) + \sin \alpha y'(a) = 0 \tag{15.2.2}$$
$$\cos \beta y(b) + \sin \beta y'(b) = 0. \tag{15.2.3}$$

We need the following important facts. The Sturm–Liouville problem defines a differential operator that is self-adjoint, and has pure point spectrum (Coddington and Levinson 1955, pp. 201 and 189), so that Theorem 15.3 applies. Thus the eigenvalues, which will be denoted by $\{\lambda_n\}$, $n \in \mathbb{N}_0$, are real; further, they are simple, countably infinite in number, have no finite point of accumulation and are bounded from below (Titchmarsh 1962, pp. 12 and 19). The eigenfunctions are orthogonal; further, they are complete and so form an orthogonal basis for $L^2(a,b)$ (Coddington and Levinson 1955, p. 198).

From the solution space of (15.2.1) a particular solution $\varphi(x, \lambda)$ can be selected which is an entire function of λ, real valued when λ is real and satisfies

$$\varphi(a, \lambda) = \sin \alpha, \qquad \varphi'(a, \lambda) = -\cos \alpha, \tag{15.2.4}$$

and another, $\psi(x, \lambda)$, which satisfies

$$\psi(b, \lambda) = \sin \beta, \qquad \psi'(b, \lambda) = -\cos \beta. \tag{15.2.5}$$

(these facts are taken from Titchmarsh (1962, pp. 6–7)). It is clear that $\varphi(x, \lambda)$ satisfies the first boundary condition (15.2.2), and that the eigenvalues $\{\lambda_n\}$ are just those values of λ for which $\varphi(x, \lambda)$ also satisfies the second boundary condition (15.2.3). Thus the eigenfunctions of the Sturm–Liouville problem are, apart from a multiplicative factor, $\{\varphi(x, \lambda_n)\}$; furthermore, they are real valued.

The Wronskian W of φ and ψ, defined by

$$W(\varphi, \psi) := \begin{vmatrix} \varphi(x, \lambda) & \psi(x, \lambda) \\ \varphi'(x, \lambda) & \psi'(x, \lambda) \end{vmatrix}, \tag{15.2.6}$$

has the following important properties.

Lemma 15.5 $W(\varphi, \psi)$ *is independent of x; it is an entire function of λ, of order $\frac{1}{2}$ and its zeros are real and simple and are located at, and only at, $\{\lambda_n\}$.*
For large k we have

$$\sqrt{\lambda_k} = \frac{k\pi}{b - a} + \mathcal{O}\left(\frac{1}{k}\right). \tag{15.2.7}$$

We also have

$$W(\lambda) := W(\varphi, \psi) = -\cos \beta \varphi(b, \lambda) - \sin \beta \varphi'(b, \lambda). \tag{15.2.8}$$

Proof All these facts except the last can be found in Titchmarsh (1962, pp. 7–11 and 19).

As to the last, it is legitimate to obtain (15.2.8) by substituting from (15.2.5) into (15.2.6), because $W(\varphi, \psi)$ is independent of x.

Theorem 15.6 *Let the notations and facts concerning the regular Sturm–Liouville problem that have been established in this section stand, and let*

$$f(\lambda) = \int_a^b u(x)\varphi(x, \lambda)\, dx, \qquad (15.2.9)$$

where $u \in L^2(a, b)$. Then we have the sampling series

$$f(\lambda) = \sum_{n=0}^{\infty} f(\lambda_n)\nu_n^{-1}\left\{ \int_a^b \varphi(x, \lambda)\varphi(x, \lambda_n)\, dx \right\}, \qquad (15.2.10)$$

where the normalizing factor ν_n is given by

$$\nu_n = \int_a^b |\varphi(x, \lambda_n)|^2\, dx.$$

The series (15.2.10) can be put into the Lagrange form

$$f(\lambda) = \sum_{n=0}^{\infty} f(\lambda_n)\frac{L(\lambda)}{L'(\lambda_n)(\lambda - \lambda_n)}, \qquad (15.2.11)$$

where L is given by (15.1.2) if 0 is not an eigenvalue, and by (15.1.3) if 0 is an eigenvalue.

The series in either of its two forms converges uniformly on compact subsets of $\mathbb{C}$.

Proof The facts we have assembled show that, with a slight notational change, $\varphi(x, \lambda)$ satisfies the requirements for the kernel in Kramer's Lemma, and since it is real for real λ we can omit complex conjugate signs. This is enough to establish (15.2.10). Now $\varphi(x, \lambda)$ is an entire function of λ so (15.2.9) shows that the same is true of f. Hence f is bounded on any compact subset of $\mathbb{C}$ and Kramer's Lemma also gives the uniform convergence.

We must now show that the series (15.2.10) and (15.2.11) are identical. We start with a standard integration formula for second order differential operators. Let one copy of (15.2.1) be written with object function $\varphi(x, \lambda)$ and a second with $\varphi(x, \lambda_n)$; let the first be multiplied by $\varphi(x, \lambda_n)$, the second by $\varphi(x, \lambda)$ and the results subtracted. This yields

$$(\lambda - \lambda_n)\varphi(x, \lambda)\varphi(x, \lambda_n) = \left(\varphi(x, \lambda)\varphi'(x, \lambda_n) - \varphi'(x, \lambda)\varphi(x, \lambda_n)\right)',$$

so that an integration gives

$$(\lambda - \lambda_n)\int_a^b \varphi(x, \lambda)\varphi(x, \lambda_n)\, dx = \varphi(b, \lambda)\varphi'(b, \lambda_n) - \varphi'(b, \lambda)\varphi(b, \lambda_n). \tag{15.2.12}$$

The integrated term vanishes at a because of the boundary condition (15.2.4). Now if $\sin \beta \neq 0$ we have, from (15.2.8),

$$
\begin{aligned}
W(\lambda)\varphi(b, \lambda_n) &= -\sin \beta \varphi(b, \lambda_n)\varphi'(b, \lambda) - \cos \beta \varphi(b, \lambda_n)\varphi(b, \lambda) \\
&= \sin \beta \left\{ \varphi(b, \lambda)\varphi'(b, \lambda_n) - \varphi'(b, \lambda)\varphi(b, \lambda_n) \right\},
\end{aligned}
$$

using (15.2.3). Hence (15.2.12) can be written

$$
\int_a^b \varphi(x, \lambda)\varphi(x, \lambda_n)\, dx = \frac{W(\lambda)}{\lambda - \lambda_n}\frac{\varphi(b, \lambda_n)}{\sin \beta}.
$$

Furthermore, taking the limit as $\lambda \to \lambda_n$, we obtain

$$
\nu_n = W'(\lambda_n)\frac{\varphi(b, \lambda_n)}{\sin \beta}.
$$

Now substitution of these quantities into (15.2.10) gives

$$
f(\lambda) = \sum_{n=0}^{\infty} f(\lambda_n)\frac{W(\lambda)}{W'(\lambda)(\lambda - \lambda_n)}. \tag{15.2.13}
$$

On the other hand, if $\sin \beta = 0$ we have, from (15.2.8),

$$
W(\lambda)\varphi'(b, \lambda_n) = -\cos \beta \varphi(b, \lambda)\varphi'(b, \lambda_n).
$$

Then using (15.2.3) we can write (15.2.12) in the form

$$
\int_a^b \varphi(x, \lambda)\varphi(x, \lambda_n)\, dx = -\frac{W(\lambda)}{\lambda - \lambda_n}\frac{\varphi'(b, \lambda_n)}{\cos \beta},
$$

and proceed to (15.2.13) as before.

Finally, because $W(\lambda)$ is an entire function of order $\frac{1}{2}$ with zeros at, and only at, $\{\lambda_n\}$, Hadamard's Factorization Theorem gives

$$
W(\lambda) = CL(\lambda),
$$

where $L(\lambda)$ is given by (15.1.2) or (15.1.3). Now (15.2.13) becomes

$$
f(\lambda) = \sum_{n=0}^{\infty} f(\lambda_n)\frac{L(\lambda)}{L'(\lambda_n)(\lambda - \lambda_n)},
$$

and the proof is complete.

Example 15.7 We return to Example 15.1 part (a) and consider the following Sturm–Liouville problem:

$$-y'' = \lambda y, \tag{15.2.14}$$
$$y'(0) = 0, \tag{15.2.15}$$
$$y'(\pi) = 0. \tag{15.2.16}$$

A fundamental system for (15.2.14) is $\{\sin x\sqrt{\lambda}, \cos x\sqrt{\lambda}\}$, from which we select $\varphi(x, \lambda) = \cos x\sqrt{\lambda}$ to satisfy (15.2.14) and (15.2.15). To satisfy (15.2.16), λ must satisfy

$$\sqrt{\lambda}\sin x\sqrt{\lambda} = 0.$$

Hence the eigenvalues are $\lambda_n = n^2, n \in \mathbb{N}_0$. Further, a straightforward calculation gives

$$\int_0^\pi \cos x\sqrt{\lambda}\cos xn\, dx = \frac{(-1)^n\sqrt{\lambda}\sin \pi\sqrt{\lambda}}{(\lambda - n^2)}.$$

Also we have $\nu_0 = \pi$ and $\nu_n = \pi/2$. Then we find from Theorem 15.6 that if

$$f(\lambda) = \int_0^\pi u(x)\cos x\sqrt{\lambda}\, dx, \tag{15.2.17}$$

then

$$f(\lambda) = f(0)\frac{\sin \pi\sqrt{\lambda}}{\pi\sqrt{\lambda}} + \frac{2}{\pi}\sum_{n=1}^\infty f(n^2)\frac{(-1)^n\sqrt{\lambda}\sin \pi\sqrt{\lambda}}{(\lambda - n^2)}. \tag{15.2.18}$$

It is noticeable from (15.2.17) that $f(\lambda)$ depends explicitly only on $\sqrt{\lambda}$, and it is in some ways more natural to put $\lambda = t^2$ and re-write this result in the following form. Let

$$g(t) = \int_0^\pi u(x)\cos xt\, dx,$$

for some $u \in L^2(0, \pi)$. Then

$$g(t) = g(0)\frac{\sin \pi t}{\pi t} + \frac{2}{\pi}\sum_{n=1}^\infty g(n)\frac{(-1)^n t\sin \pi t}{t^2 - n^2}.$$

This recovers the "even form" of the cardinal series, the first of its variants that we noted in Chapter 1.

The example we have just looked at raises the question of whether it is always the case that $f(\lambda)$ in (15.2.9) depends explicitly only on $\sqrt{\lambda}$. This does turn out to be the case, as we shall see in Corollary 15.8. The main reason for re-writing Theorem 15.6 in this form is to be found in the Corollary 15.9 to follow, where a connection with the Nyquist–Landau sampling rate is made.

We need some further facts from the theory of Sturm–Liouville problems (they are taken from Zayed et al. 1990). If we put $\lambda = t^2$ and $\lambda_n = t_n^2$ there is a solution $K(x,t)$ of (15.2.1) and (15.2.2) having the following properties: it is an entire function of t, and as a function of x it is real valued whenever t^2 is real, and it does not vanish identically for any value of t. For example, we have $\varphi(x, \lambda_n) = K(x, t_n)$. We now have:

Corollary 15.8 *Let*

$$g(t) = \int_a^b u(x)K(x,t)\,dx \qquad (15.2.19)$$

for some $u \in L^2(a,b)$. Then

$$g(t) = \sum_{n=0}^{\infty} g(t_n) \frac{L(t^2)}{L'(t_n^2)(t^2 - t_n^2)}.$$

Corollary 15.9 *The sample points $\{t_n\}$ in Corollary 15.8 have density $(b - a)/2\pi$.*

Proof Now, from Definition 10.3, the density $D(t_n)$ of $\{t_n\}$ is

$$D(t_n) = \lim_{r \to \infty} \frac{\#\{t_k : t_k \in [-r, r]\}}{2r}.$$

But from (15.2.7) we have

$$t_k = \frac{k\pi}{b-a} + \mathcal{O}\left(\frac{1}{k}\right), \qquad k \to \infty.$$

Hence, for all sufficiently large k, there are k members of $\{t_n\}$ in the interval $[-(k+\tfrac{1}{2})\pi/(b-a), (k+\tfrac{1}{2})\pi/(b-a)]$. Hence

$$D(t_n) = \lim_{k \to \infty} \frac{k}{2(k+\tfrac{1}{2})\pi/(b-a)} = \frac{b-a}{2\pi} \lim_{k \to \infty} \frac{k}{k+\tfrac{1}{2}} = \frac{b-a}{2\pi}.$$

This density is analogous to the Nyquist–Landau sample point density of Definition 1.4, and is, perhaps, what one would expect. However, it remains a matter of conjecture as to whether there is an analogue of Landau's Theorem, §17.5, guaranteeing that this is the *minimum* density at which a function of the form (15.2.19) can be sampled and reconstructed.

$$* \quad * \quad *$$

Up to this point it has been assumed that a Sturm–Liouville problem has been given *a priori*, and we have seen that its eigenvalues then determine the sample points in the associated sampling series. However, it may be desirable to work from a given set of sample points, and in this case we need to consider the inversion of the Sturm–Liouville problem. This consists of asking for conditions under which a real sequence (λ_n) can be prescribed for which there is a regular Sturm–Liouville eigenvalue problem having eigenvalues $\{\lambda_n\}$. When this problem is solvable for a regular Sturm–Liouville eigenvalue problem, Theorem 15.6 shows how to construct a sampling series with sampling points $\{\lambda_n\}$, and gives the connection with Lagrange interpolation. This procedure gives us some freedom in the choice of sampling points, but it should be noted that there is no more freedom once this choice has been made, because the class of functions to which the series applies is determined by the resulting Sturm–Liouville problem.

The method is based on the following theorem of Levitan and Gasymov, and needs no further explanation (remarks and references can be found in Zayed *et al.* (1990).

Theorem 15.10 *Let* (λ_n), $n \in \mathbb{N}$, *be a sequence of distinct real positive numbers, and let* (τ_n), $n \in \mathbb{N}$ *and* (ρ_n), $n \in \mathbb{N}$ *both belong to* l^2. *Let a, b and c be constants, and suppose further that*

$$\lambda_n = \frac{n}{2} + \frac{a}{n} + \frac{b}{n^3} + \frac{\tau_n}{n^3}, \qquad n \in \mathbb{N},$$

and that in the sequence

$$r_n = \frac{1}{\pi} + \frac{c}{n^2} + \frac{\rho_n}{n^3}$$

each r_n *is positive. Then there exists a regular Sturm–Liouville eigenvalue problem having eigenvalues* $\{\lambda_n\}$, *and for which* $\{r_n\}$ *are the normalizing factors for the eigenfunctions.*

15.3 Further types of eigenvalue problem — some examples

The research literature on sampling theory associated with eigenvalue problems has progressed far beyond the regular Sturm–Liouville problem. In particular, there are now many results on sampling associated with singular eigenvalue problems, a class which Titchmarsh (1962, preface) points out "seems to contain all the most interesting examples". Regrettably, it would take us much too far afield to take up the singular theory here.

In this section some examples are discussed which stay within the realms of regularity but in which some of the rather stringent restrictions of the regular Sturm–Liouville problem are relaxed. For example, Butzer and Schöttler (1994) have shown that Theorem 15.6 has an analogue for *n*th order regular self-adjoint eigenvalue problems with simple spectrum, and in Example 15.11 one of their fourth order cases is presented (further interesting remarks about the example can be found in their article).

Two more examples are presented that are not part of any currently known theoretical framework. In Example 15.12 we look at a problem with "double" eigenvalues, and in Example 15.13 we look at a non-self-adjoint problem. An interesting open problem here is to supply general theories which would account for examples of these two types.

Example 15.11 The fourth order eigenvalue problem consisting of the equation[1]

$$y^{(iv)} = \lambda y \qquad (15.3.1)$$

on $[0, \pi]$, together with the boundary conditions

$$
\begin{aligned}
y'(0) &= 0 \\
y'''(0) &= 0 \\
y'(\pi) &= 0 \\
y'''(\pi) &= 0
\end{aligned}
\qquad (15.3.2)
$$

defines a self-adjoint differential operator,[2] densely defined on $L^2(0, \pi)$ and with pure point spectrum. Hence Theorem 15.3 applies. As in the previous theory, a kernel must be selected from the solution space of (15.3.1), and this time it must satisfy not only (15.3.1) but also the first three boundary conditions of (15.3.2). As before, it is convenient to state our theorem in terms of t, where this time we put $\lambda = t^4$. It must be emphasized that the selection of a suitable kernel is a *very* delicate matter. Thus, let us choose

$$\varphi(x, t) := \frac{1}{t} \left(\cosh xt \sin \pi t + \cos xt \sinh \pi t \right),$$

and note that the factor $1/t$ is indispensible here in order to keep the kernel (which would have the correct properties in other respects) from vanishing identically when $t = 0$. In fact, we find that $\varphi(x, 0) \equiv 2\pi$, and it is an easy matter to verify that this choice for $\varphi(x, t)$ satisfies 15.3.1 and the first three boundary conditions.

Now the point spectrum is determined from the fourth boundary condition; this reduces to

$$\sinh \pi t \sin \pi t = 0.$$

---

[1] This equation was discovered independently by Daniel Bernoulli and Leonhard Euler, sometime before May 1735, when it appeared in correspondence between them; at this time they were exchanging ideas on the displacements of elastic bodies. Further interesting information can be found by reading C. Truesdell in The rational mechanics of flexible or elastic bodies 1638–1788; introduction to Leonhardi Euleri Opera Omnia Vol. X et XI Seriei Secundae, Orell Füssli, Turici (Zurich), 1960; particularly section 24, pp. 165–170.

[2] This operator is encountered in the vibration analysis of horizontal beams with uniform cross-section. The boundary conditions are of the "guided support" type, where the end-points of the beam are clamped to rollers which allow free movement in the vertical direction only. Hence they are not free to rotate (this is expressed by the "dynamic boundary conditions" $y'(0) = y'(\pi) = 0$), and there is no shearing force at the end-points (this is expressed by the "static boundary conditions" $y'''(0) = y'''(\pi) = 0$). Source: Egor P. Popov (1968), *Introduction to mechanics of solids*, Prentice Hall, p. 386.

So 0 is an eigenvalue, and so are the positive integers; thus, the eigenvalues are $\{n\}$, $n \in \mathbb{N}_0$. The corresponding eigenfunctions consist of the constant function 2π together with $\{(1/n) \sinh \pi n \cos xn\}$. It is now a routine matter to calculate $S_0(t)$ and $S_n(t)$ using the formulae in Kramer's Lemma.

Altogether we have the following sampling series. Let

$$g(t) = \frac{1}{t} \int_0^\pi u(x)\{\cosh xt \sin \pi t + \cos xt \sinh \pi t\}\, dx$$

for some $u \in L^2(0, \pi)$. Then

$$g(t) = g(0)\frac{\sinh \pi t}{\pi t}\frac{\sin \pi t}{\pi t} + \frac{4}{\pi}\sum_{n=1}^\infty g(n)\frac{(-1)t^2 n \sinh \pi t \sin \pi t}{\sinh \pi n(t^4 - n^4)}.$$

This can be put into the Lagrange form by using the product representation

$$\sinh \pi t \sin \pi t = \pi^2 t^2 \prod_{k=1}^\infty \left(1 + \frac{t^2}{k^2}\right)\prod_{k=1}^\infty \left(1 - \frac{t^2}{k^2}\right) = \pi^2 t^2 \prod_{k=1}^\infty \left(1 - \frac{t^4}{k^4}\right).$$

Example 15.12 We consider again the equation

$$-y'' = \lambda y,$$

this time with periodic boundary conditions

$$y(-\pi) - y(\pi) = 0$$
$$y'(-\pi) - y'(\pi) = 0.$$

A fundamental system of solutions is $\{\cos x\sqrt{\lambda}, (1/\sqrt{\lambda})\sin x\sqrt{\lambda}\}$. Clearly, 0 is a simple eigenvalue with the constant function 1 as corresponding (normalized) eigenfunction, since the second member of the fundamental system contributes the solution $y = x$ in this case, and this does not satisfy the first boundary condition. Now $\cos x\sqrt{\lambda}$ satisfies the first boundary condition, and the second if $\sqrt{\lambda} = n$, $n \in \mathbb{N}$. Similarly, $\sin x\sqrt{\lambda}$ satisfies the second boundary condition, and the first again if $\sqrt{\lambda} = n$.

Hence $\lambda = n^2$ is a double eigenvalue with corresponding eigenspace spanned by $\{\cos nx, \sin nx\}$. Now $L_e^2(-\pi, \pi)$, consisting of the even members of $L^2(-\pi, \pi)$, is spanned by the orthogonal set $\{1, \cos xn\}$. As in Example 15.7 we can now proceed to the "even" form of the cardinal series for f_e, an even member of PW.

Similarly, $L_o^2(-\pi, \pi)$, consisting of the odd members of $L^2(-\pi, \pi)$, is spanned by the orthogonal set $\{\sin nx\}$, $n \in \mathbb{N}$, and similarly one can obtain the "odd" form of the cardinal series (1.1.9) for f_o, an odd member of PW. Since

$$L^2(-\pi, \pi) = L_e^2(-\pi, \pi) \oplus L_o^2(-\pi, \pi),$$

the ordinary cardinal series is recovered from this "double" eigenvalue problem by taking $f = f_e + f_o \in PW$.

Example 15.13 We return to Example 15.1 part (b) and consider again the equation

$$-y'' = \lambda y,$$

with mixed boundary conditions

$$\begin{aligned} y'(-\pi) &= 0 \\ y(-\pi) + y'(\pi) &= 0. \end{aligned} \tag{15.3.3}$$

This problem defines a differential operator T that is not of the Sturm–Liouville type. It is not self-adjoint, as the reader has been invited to check; the reader may also like to check that the adjoint operator T^* is given by the eigenvalue problem

$$\begin{aligned} -y'' &= \lambda y \\ y'(\pi) &= 0 \\ y'(-\pi) - y(\pi) &= 0. \end{aligned} \tag{15.3.4}$$

To obtain a sampling theorem we need the biorthogonal form of Kramer's Lemma; two kernels must be selected, one from T and one from T^*. As before we put $\lambda = t^2$. Now for T we choose $K(x,t) = \cos[(x + \pi)t]$, which satisfies the equation and the first boundary condition of (15.3.3). Hence the spectrum is determined from the second boundary condition; thus, the eigenvalues $\{t_n\}$ are the positive roots of $\sin 2\pi t = 1/t$ and T has (un-normalized) eigenfunctions $\{K(x, t_n)\} = \{\cos[(x + \pi)t_n]\}$. That these functions form a Riesz basis for $L^2(-\pi, \pi)$ follows from a theorem of Mihailov (1962).

For T^* we choose $K^*(x,t) = \cos[(x - \pi)t]$; the spectrum is the same as that of T, and the (un-normalized) eigenfunctions are $K^*(x, t_n) = \cos[(x - \pi)t_n]$.

With this information we can calculate $S_n(t)$. We shall need the integral

$$\int_{-\pi}^{\pi} \cos[(x + \pi)t] \cos[(x - \pi)t_n]\, dx = \frac{t \sin 2\pi t - 1}{t^2 - t_n^2}$$

and ν_n, which is obtained by taking $\lim_{t \to t_n}$ in the previous integral. We obtain

$$\nu_n = \frac{\sin 2\pi t_n}{2t_n} + \pi \cos 2\pi t_n.$$

Altogether we have the following sampling expansion. Let

$$f(t) = \int_{-\pi}^{\pi} u(x) \cos[(x + \pi)t]\, dx$$

for some $u \in L^2(-\pi, \pi)$. Then

$$f(t) = \sum_{n=1}^{\infty} f(t_n) \frac{t \sin 2\pi t - 1}{\nu_n \pi (t^2 - t_n^2)},$$

where $\{t_n\}$ are the positive roots of $t \sin 2\pi t = 1$, and

$$\nu_n = \int_{-\pi}^{\pi} \cos[(x + \pi)t_n] \cos[(x - \pi)t_n] \, dx.$$

The dual part of the biorthogonal form of Kramer's Lemma leads to exactly the same expansion.

It is also interesting to note that this expansion can be put into the Lagrange form, by putting $L(\lambda) = \sqrt{\lambda} \sin 2\pi \sqrt{\lambda} - 1$. Then

$$L'(t_n^2) = \pi \cos 2\pi t_n + \frac{\sin 2\pi t_n}{2t_n} = \nu_n.$$

On putting $\lambda = t^2$ we obtain the reconstruction functions as before. We might also note that $L(\lambda)$ is an entire function of order $\frac{1}{2}$, so that Hadamard's Factorization Theorem applies, and it is represented by the canonical product taken over its zeros which are just the eigenvalues $\{t_n\}$.

A more advanced treatment of sampling and eigenvalue problems is planned for inclusion in H–S. Meanwhile, further developments can be found in, for example, Zayed (1993, Chs 4–6), Butzer and Schöttler (1993), Annaby (to appear), and Everitt *et al.* (1994) .

16

CAMPBELL'S GENERALIZED SAMPLING THEOREM

In this chapter we are going to study a series expansion of the form

$$f(t) = \sum_{n \in \mathbb{Z}} L_n(f) S_n(t), \tag{16.0.1}$$

in which $\{S_n\}$ is a set of expansion functions and $\{L_n\}$ a set of associated coefficient functionals of a particular kind, which will be defined shortly (in (16.1.1) and (16.1.3) below). This series contains the ordinary cardinal series

$$f(t) = \sum_{n \in \mathbb{Z}} f(\frac{n}{w}) \operatorname{sinc}(wt - n)$$

for a Paley–Wiener function f as a special case, as we shall see in §16.2. The main questions we want to ask here are concerned with the extent to which properties of the cardinal series persist in this more general form.

Two properties are of particular interest. First, since the coefficients in the cardinal series are measurements of f which are, ideally, completely localized in time, it will be of interest to ask if this is also true of L_n. We shall study some cases in which this is true, but in a less clear-cut way; in fact, in such a way that L_n depends essentially on just a few samples, especially if $|n|$ is small.

Second, we shall want to know if there is any interdependence between the coefficient functionals $\{L_n\}$. For example, if f belongs to some space for which $\{S_n\}$ is a basis, as in the Paley–Wiener case, then there can be no dependency relations among the coefficient functionals. In the same vein, the property of being statistically uncorrelated is desirable for $\{L_n\}$.

16.1   L.L. Campbell's generalization of the sampling theorem

Essential ingredients of Campbell's generalization of the sampling theorem are a probability distribution $\varphi(x)$ and its cumulative distribution $\alpha(x)$, although we shall need very little in the way of probability theory to discuss the series.

Let φ be measurable and non-negative over $\mathbb{R}$, and such that $\int_{\mathbb{R}} \varphi = 1$. Let

$$\alpha(x) = \int_{-\infty}^{x} \varphi(u) \, du.$$

Let $E \subseteq \mathbb{R}$ be the set $\{x : \varphi(x) > 0\}$. Now α restricted to E is increasing so that we can define an inverse function $\beta = \alpha^{-1}$, so that $\beta(\alpha(x)) = x$, $x \in E$.

The series expansion that we are going to discuss is based on the the Fourier series for the "kernel" $e^{i\beta(u)t}$.

Lemma 16.1 *We have*

$$e^{i\beta(u)t} = \sum_{n \in \mathbb{Z}} S_n(t)e^{2\pi inu}$$

with convergence in the norm of $L^2(0,1)$, where

$$S_n(t) = \int_{\mathbb{R}} \varphi(v)e^{-2\pi in\alpha(v)}e^{itv}\,dv. \qquad (16.1.1)$$

Proof The coefficients $S_n(t)$ are given by

$$\int_0^1 e^{i\beta(x)t}e^{-2\pi inx}\,dx.$$

After the change of variable $x = \alpha(v)$ this gives (16.1.1).

Next, let ρ be a "weight" function which is non-negative, belongs to $L(\mathbb{R})$ and vanishes together with φ; that is, $\rho(x) = 0$ when $x \notin E$. In many applications ρ is taken to be the spectral density of a stationary random process for which a series expansion is sought.

Let F be the class of functions f such that

$$f(t) = \int_{\mathbb{R}} g(x)e^{ixt}\rho(x)\,dx$$

for some $g \in L^2(\mathbb{R}, \rho\,dx)$. The following theorem and its proof are strongly reminiscent of Kramer's Lemma (§8.1), and can be considered as a companion form of it.

Theorem 16.2 *Let $\rho(x)/\varphi(x)$ be bounded for $x \in E$. Then, for every $f \in F$, we have*

$$f(t) = \sum_{n \in \mathbb{Z}} L_n(f)S_n(t), \qquad (16.1.2)$$

where S_n is given by (16.1.1), and where

$$L_n(f) = \int_{\mathbb{R}} g(x)e^{2\pi in\alpha(x)}\rho(x)\,dx, \qquad (16.1.3)$$

the convergence being pointwise over $\mathbb{R}$.

Proof

$$\left| f(t) - \sum_N L_n(f) S_n(t) \right|^2$$

$$= \left| \int_{\mathbb{R}} \left\{ e^{ixt} - \sum_N e^{2\pi in\alpha(x)} S_n(t) \right\} g(x)\rho(x)\, dx \right|^2$$

$$\leq \|g\| \int_{\mathbb{R}} \left| e^{ixt} - \sum_N e^{2\pi in\alpha(x)} S_n(t) \right|^2 \rho(x)\, dx. \qquad (16.1.4)$$

After making the change of variable $x = \beta(u)$, (16.1.4) becomes

$$\|g\| \int_0^1 \left| e^{i\beta(u)t} - \sum_N e^{2\pi inu} S_n(t) \right|^2 \frac{\rho(\beta(u))}{\varphi(\beta(u))}\, du.$$

Because ρ/φ is bounded, this expresssion does not exceed

$$(\text{constant}) \|g\| \int_0^1 \left| e^{i\beta(u)t} - \sum_N e^{2\pi inu} S_n(t) \right|^2 du,$$

and this last expression is vanishingly small for large N, by Lemma 16.1.

Investigations in this area have shown that when ρ is a constant multiple of φ (in which case ρ/φ is of course bounded) the coefficients $\{L_n\}$ are uncorrelated random variables, so this an important case for series expansions of random processes.

Let us also note at this stage that from (16.1.3) we always have

$$L_0(f) = \int_{\mathbb{R}} g(x)\rho(x)\, dx = f(0).$$

That is, the zeroth coefficient is always a sample of f, and as such is of course completely localized in time. This can be approximately true for other coefficients L_n. Suppose, for example, that α is close to being linear, say $\alpha(x)$ is approximately equal to $hx + c$. Then, from the calculation

$$\left| L_n(f) - e^{2\pi inc} f(2\pi nh) \right| = \left| \int_{\mathbb{R}} g(x)[e^{2\pi in\alpha(x)} - e^{2\pi in(hx+c)}]\rho(x)\, dx \right|$$

$$\leq \|g\| \, \left\| e^{2\pi in\alpha(x)} - e^{2\pi in(hx+c)} \right\|,$$

in which each norm is that of $L^2(\mathbb{R}, \rho\, dx)$, we find that if the linear approximation for α is reasonably accurate, then L_n is nearly determined by the sample $f(2\pi nh)$, at least for small $|n|$.

16.2 Band-limited functions

We can recover the ordinary cardinal series for Paley–Wiener functions by taking $\varphi(x) = X_{\pi w}(x)/2\pi w$ and $\rho(x) = c\varphi(x)$, where c is a constant. With these specializations it is a straightforward matter to calculate

$$\alpha(x) = \frac{1}{2\pi w}(x + \pi w), \qquad |x| \le \pi w,$$

$$L_n(f) = \frac{(-1)^n c}{2\pi w} \int_{-\pi w}^{\pi w} g(x) e^{inx/w}\, dx,$$

$$f(t) = \frac{c}{2\pi w} \int_{-\pi w}^{\pi w} g(x) e^{ixt}\, dx, \qquad g \in L^2(-\pi w, \pi w).$$

Furthermore,

$$S_n(t) = \frac{1}{2\pi w} \int_{-\pi w}^{\pi w} e^{-\pi i n(1 + x/\pi w)} e^{ixt}\, dx = (-1)^n \operatorname{sinc}(wt - n).$$

Hence Theorem 16.2 gives the cardinal series representation for $f \in PW_{\pi w}$.

For applications in detection theory the cardinal series is particularly useful when a signal is to be detected in the presence of white noise which is band-limited to $[-\pi w, \pi w]$. This is because sample values of white noise at regularly placed time points are uncorrelated, and statistical difficulties are minimized. The more general series (16.0.1) will be useful in this context if one can choose φ in such a way as to yield uncorrelated coefficients L_n when coloured noise of spectral density $\rho(x)$ is expanded. Experience shows that a good choice is to make φ proportional to ρ.

In practice this could be difficult, however, because it may be necessary to work with a ρ that is given *a priori* in a complicated, or even partially incomplete form. It seems to be a reasonable strategy, then, to stay within the realms of band-limited functions by choosing φ to have support on $[-\pi w, \pi w]$, but to make it more general than the choice that gave us the cardinal series by adding cosine terms. For example, let us investigate the consequences of taking

$$\varphi(x) = \left\{ \frac{1}{2\pi w} + \epsilon \cos\left(\theta + \frac{x}{w}\right) \right\} \chi_{\pi w}(x).$$

From the formula for α we calculate

$$\alpha(x) = \tfrac{1}{2} + \frac{x}{2\pi w} + \epsilon w \left(\sin(\theta + \frac{x}{w}) + \sin\theta \right),$$

and from (16.1.3) we obtain

$$L_n(f) = (-1)^n e^{2\pi i n \epsilon w \sin\theta} \int_{\mathbb{R}} e^{inx/w} e^{2\pi i n \epsilon w \sin(\theta + x/w)} g(x) \rho(x)\, dx.$$

It is now necessary to introduce one of the generating relations for Bessel functions J_m of integer order m,

$$e^{ia\sin\eta} = \sum_{m\in\mathbb{Z}} J_m(a) e^{im\eta}.$$

This formula and other relevant facts can be found in Whittaker and Watson (1962, p. 357). Then

$$L_n(f) = e^{\pi i n(1 + 2\epsilon w \sin\theta)} \sum_{m \in \mathbb{Z}} J_m(2\pi n\epsilon w)e^{im\theta} f\left(\frac{n+m}{w}\right).$$

This shows that the action of L_n on f is to form an infinite linear combination of samples of f. However, if ϵ is small only those terms containing samples close to $f(n/w)$ will contribute much to L_n, and in any case L_n will be dominated by the term containing $f(n/w)$.

More generality would be achieved by adding further cosine terms to φ. Thus, the next stage would be to take

$$\varphi(x) = \left\{\frac{1}{2\pi w} + \epsilon_1 \cos\left(\theta_1 + \frac{x}{w}\right) + \epsilon_2 \cos\left(\theta_2 + \frac{2x}{w}\right)\right\} \chi_{\pi w}(x).$$

The analysis would proceed in a similar way to that in the previous case. Then L_n would still be a linear combination of samples of f, but of a much more complicated kind. Nevertheless, when ϵ_1 and ϵ_2 are small L_n would again be dominated by a few samples near to $f(n/w)$.

Complexity increases with the addition of yet further cosine terms of course, but the feature that L_n is dominated by samples of f at sample points close to n/w remains.

16.3   Non-band-limited functions — an example

We end this chapter with an example to illustrate what can happen when f is not band-limited.

Example 16.3 Take

$$\varphi(x) = \frac{a}{\pi} \frac{1}{x^2 + a^2}, \qquad x \in \mathbb{R}.$$

Then we calculate

$$\alpha(x) = \frac{a}{\pi} \int_{-\infty}^{x} \frac{1}{a^2 + v^2} \, dv = \frac{1}{\pi} \arctan\frac{x}{a} + \frac{1}{2}.$$

Also it is straightforward to calculate

$$e^{2\pi i n\alpha(x)} = \left(\frac{ix+a}{ix-a}\right)^n, \qquad n \in \mathbb{Z},$$

and when

$$f(t) = \int_{\mathbb{R}} g(x)e^{ixt}\rho(x) \, dx$$

for some $g \in L^2(\mathbb{R}, \rho \, dx)$, we have

$$L_n(f) = \int_{\mathbb{R}} \left(\frac{ix + a}{ix - a} \right)^n g(x)\rho(x) \, dx. \tag{16.3.1}$$

If $\rho(x)$ is chosen to be a constant multiple of $\varphi(x) = (x^2 + a^2)^{-1}$, the coefficients $\{L_n\}$ will be, as already noted, statistically uncorrelated random variables. As to the possibility of time localization, let us examine the form of L_n more closely; here it should be noted that the argument does not require a specific choice for ρ. We have

$$\left(\frac{ix + a}{ix - a} \right)^n = \left(1 + \frac{2a}{ix - a} \right)^n = 1 + \sum_{k=1}^{n} \binom{n}{k} \frac{(2a)^k}{(ix - a)^k}.$$

Now the integral

$$\int_0^{\infty} e^{(ix-a)t} t^{k-1} \, dt = \frac{(-1)^k (k-1)!}{(ix - a)^k}$$

is used to re-write this last expression in the form

$$1 + \sum_{k=1}^{n} \binom{n}{k} \frac{(2a)^k (-1)^k}{(k-1)!} \int_0^{\infty} e^{(ix-a)t} t^{k-1} \, dt,$$

and substitution into (16.3.1) gives, for $n > 0$,

$$L_n(f) = \int_{\mathbb{R}} g(x) \left\{ 1 + \sum_{k=1}^{n} \binom{n}{k} \frac{(-2a)^k}{(k-1)!} \int_0^{\infty} e^{(ix-a)t} t^{k-1} \, dt \right\} \rho(x) \, dx$$

$$= f(0) + \sum_{k=1}^{n} \binom{n}{k} (-2)^k l_k(f),$$

where we have put

$$l_k(f) = \frac{a^k}{(k-1)!} \int_0^{\infty} f(t) e^{-at} t^{k-1} \, dt.$$

Now since

$$\frac{a^k}{(k-1)!} \int_0^{\infty} e^{-at} t^{k-1} \, dt = 1$$

we can interpret $l_k(f)$ to be the expected value of f with respect to the probability distribution $a^k e^{-at} t^{k-1}/(k-1)!$, which has the mean

$$\frac{a^k}{(k-1)!} \int_0^{\infty} f(t) e^{-at} t^k \, dt = \frac{k}{a}.$$

It appears that $L_n(f)$ is a linear combination of the quantities $l_k(f)$, $k = 1, \ldots n$, each of which is the expected value of a distribution with mean k/a. We can think of L_n as being time localized in that it is a combination of the samples $\{f(k/a)\}$ in this special sense.

This chapter is based on the work of Campbell (1985), which contains several further interesting results and references.

PROBLEMS

16.1 In Example 16.3 find L_n for $n < 0$, and show that there is a separation of past effects from future effects in the series representation (16.0.1) (the present time being $t = 0$). In particular, show that, for $n > 0$, S_n vanishes on the negative half of $\mathbb{R}$, and on the positive half can be expressed in terms of Laguerre polynomials.

16.2 Find the variance of the distribution $a^k e^{-at} t^k / (k-1)!$, and use it to show that time localization in Example 16.3 cannot be said to persist for large $|n|$.

17

MODELLING, UNCERTAINTY AND STABLE SAMPLING

The main goal of this chapter is to reach Landau's theorem (Theorem 17.9) on the minimum stable sampling rate. Along the way we shall encounter several interesting aspects of signal modelling.

The proof of Theorem 17.9, as presented in §17.5, is not fully complete and rests in some parts on heuristic arguments. A completely detailed proof would be too long for inclusion here, but we shall go as far as we can in order to get a feel for this important theorem. The complete proof, in the multi-dimensional case, can be found in Landau (1967b).

17.1 Remarks on signal modelling

"It is easy to argue that real signals must be band-limited. It is also easy to argue that they cannot be so." (Slepian 1976, p. 292).

Slepian deliberately coined this capricious statement to highlight a paradox of signal modelling that one can fall into rather easily. The invitation to put matters into a more satisfactory light is clear, and this will occupy much of the present section. Another paradox, this time involving prediction and rather more subtle than the first one, will also try to frustrate our efforts; but hopefully we shall be equal to the task.

The discussions are based largely on expository writings of Landau, Pollak and Slepian (references are given in context), and a few more remarks have been added here including a "thought experiment". This will help to set the scene for the uncertainty principle of signal analysis, treated in §17.3, and will allow us to look at some other important points about signal modelling, in particular the effectiveness of modelling signals using Paley–Wiener functions.

The argument that a real signal must be band-limited is based on the simple observation that band-limitation is a natural physical constraint. This is because it is reasonable to assume that physical materials do not have the capacity for arbitrarily rapid vibration, so that a signal in the real world must be expected to have bounded frequency content.

Another reasonable expectation about real signals is that they can have only a finite amount of energy, and it will be recalled that these two properties — band-limitation and finite energy — are just those that characterize Paley–Wiener functions.

The argument on the other side, that a real signal cannot be band-limited, comes from a property of Paley–Wiener functions that we saw in §1.3; they cannot be of finite duration. But real signals start and stop; they do not last for ever as Paley–Wiener functions are required to do, so the finite energy property together with time-limitation prevents a real signal from being band-limited.

Now, we do not normally make any great distinction between a "real signal" and a "mathematical model" of a real signal, because difficulties do not normally arise if both are treated interchangeably. But now, confronted with our paradox we shall be forced to recognize that there is an important distinction to be made, and we must take a careful look at this before the paradox can be resolved.

First, what shall we mean by a *real signal*? Examples of what this is likely to mean are not difficult to find. The sound waves involved in human speech, music or bird song, for example; the electromagnetic waves associated with television or radio; the radiation by which we receive heat and light from the sun — all are examples of real signals encountered in our daily lives, and of course there are many more.

Looking towards the modelling aspect, we would naturally want to endow our models with properties that reflect our perceptions of reality. For example, our real signals certainly do appear to start and to stop; they seem to vibrate within a certain bounded range of frequencies; they appear to have finite energy; and the list goes on.[1]

But all of this is rather vague. We really need to be more scientific and adopt the physicists' "minimalist" frame of mind when they say that "the only reality is what we can detect". In this approach a real signal is considered to be known only in so far as we can make measurements or observations on it. If these acts of observation are time-limited, and they always are in practice of course, then we will observe a signal that is apparently time-limited; if we observe frequencies only within certain frequency bands, then we will observe a signal with the apparent property of band-limitation. However, the time duration and bandwidth, the energy, information content, entropy, continuity or differentiability of a signal are really all human artefacts which we use to describe what we observe. These are properties of the *mathematical model*. The nature of a *real signal* in our minimalist framework is to be found only in the observations made on it; its "true nature", whatever that may mean, is not the object of scientific investigation.

We can now resolve our first paradox. It is meaningless to ask if a real signal is band-limited, because real signals are just collections of observations and have no properties beyond this. We may wish to assume a Paley–Wiener function

---

[1]As long ago as 400 BC or thereabouts, Archytas (of Tarentum, 428–365 BC) suggested a connection between musical pitch and frequency of vibration, saying "Clearly swift motion produces a high-pitched sound, slow motion a low-pitched sound". Later, Euclid (of Alexandria, c. 330–c. 275 BC) wrote "Some sounds are higher pitched, being composed of more frequent and more numerous motions". These remarks are quoted by C. Truesdell, with further references, in his Prologue to The rational mechanics of flexible or elastic bodies 1638–1788; introduction to Leonhardi Euleri Opera Omnia Vol. X et XI Seriei Secundae, Orell Füssli, Turici (Zurich), 1960.

Sound associated with vibratory motion and its frequency was not a fully accepted idea until about 1600 AD, but certainly had its roots in classical antiquity. It seems that in those days people were already beginning to think about signal modelling!

as the model; then we are assuming the property of band-limitation. In other circumstances it might be appropriate to assume a time-limited model (which, because of the finite energy assumption, could not then be band-limited). In any case, band-limitation is a property of a mathematical model, not of a real signal.

We have seen how confusion can arise if one does not distinguish properly between real signals and their models. So, from now on we ought to bear in mind that "the signal is not the model".

This rule is effective in resolving a second paradox — the paradox of prediction.

"You give me your speech from yesterday and I can tell you what you are going to say tomorrow" was the example given by Pollak (1963, p. 74) (and echoed by Landau (1985 p. 202)).

First, we recall that a Paley–Wiener function can be extrapolated, that is, continued analytically into any region from any part of itself; from a line segment, for example, no matter how small. If we argue, as we did in our first paradox, that real signals are of finite energy and do not vibrate at arbitrarily high frequencies, and that consequently they seem to have all the properties of Paley–Wiener functions, we must include among these properties that of complete predictability. On the other hand, we know that real signals are not completely predictable, and herein lies our second paradox.

This paradox arises because once again we have fallen into the trap of confusing the real signal with the model. The model may have the capacity for prediction, but the real signal remains as elusive as ever. However, we can still ask why the model does not give a perfect prediction of the signal.

Pollak blames instabilities in the prediction process, saying, in effect, that discrepancies in measuring the samples can lead to arbitrarily large discrepancies in the prediction, and arbitrarily soon. The process of prediction is the process whereby the model is constructed from the data, and while the resulting model may have the capacity to predict its own future, prediction of the real signal will of course depend on how well the model tracks the signal into the future. Pollak's contention is that this may not be very far and that model and signal are destined to part company eventually. It will be instructive to look into this a little further.

The following "thought experiment" may be helpful. Its intention is to exhibit a real signal that is locally modelled very accurately by a Paley–Wiener function, but which nevertheless has virtually no capacity for prediction from this Paley–Wiener model.

Suppose that two digital recordings of a musical performance are made on compact disc, one of the first half, the other of the whole. These recordings consist of samples of the music taken during their respective periods of observation, and by implication samples that are zero at other times. In order to drive the loudspeaker, the playing equipment is designed to produce a Paley–Wiener function from these samples (by the standard techniques of digital-to-analogue conversion, which, in its basic form is called "sample and hold").

It is safe to assume that the recording and sound reproduction technology is so good that the listener can hardly distiguish between the outputs from these two Paley–Wiener models over the first half of the performance, or indeed between them and the actual performance.

If these two models were identical over the first half of the performance, then the first half would completely predict the second half, at least as far as the listener is concerned! This absurdity presents a compelling case for arguing that the two models, although close, are not in fact identical; indeed, this is not incompatible with known behaviour of Paley–Wiener functions.

The fact is that one can find an interval, even an arbitrarily long one, over which any two Paley–Wiener functions are arbitrarily close. For example, both must approach zero as their arguments become large, and so we can find a long interval where they are indistinguishable above any pre-assigned detection level; in general they will differ radically elsewhere. The two music signals are close enough to deceive the listener over the first half of the performance, but they will differ fundamentally over the second half.

Our two music signals are transient phenomena and their Paley–Wiener models evidently do an excellent job of modelling them. This is possible because, although not exactly time-limited, a Paley–Wiener function can be *nearly* time-limited in that most of its energy is concentrated on a bounded time interval; outside this interval it decays rapidly, and eventually remains below any given detection level.

The *prolate spheroidal wave functions* are examples of this kind. We shall encounter these functions in Theorem 17.5, parts 1 and 4 (where they are denoted $\{\varphi_n\}$), after which they will appear regularly in our discussions as their properties are investigated. A brief summary of these can be found in §17.4. Paley–Wiener functions are by no means innappropriate for modelling transient phenomena.

The fact that a Paley–Wiener function can be "near" to a time limited function has far-reaching consequences. It leads directly to the uncertainty principle of signal analysis, as we shall see in §17.3., and the same ideas then lead up to Theorem 17.9 in the final section.

Our thought experiment concludes with one further observation. Let us return to the model of the first half of our musical performance, and ask what happens if we adjoin samples from the second half to it. Eventually we should obtain the model of the whole, but we just saw that the two models cannot be identical over the first half. Why does there seem to be a discrepancy here?

The fact is that the adjunction of even one further sample changes the whole model. This is because a sample that was zero has been replaced with one that is not zero and so a new Paley–Wiener function results. The correct model of the whole would indeed be reached, as one would expect, by successively adjoining all the remaining samples.

This concludes our thought experiment, and in it we have seen that a Paley–Wiener model does not necessarily project much information about the signal it models beyond the interval of observation. In fact the model of the first half of our musical performance tells the listener not blessed with super-human hearing to expect nothing but silence from the second half.

The signal is certainly not the model!

It may be said that if we are careful always to make a proper distinction between real signals and those mathematical functions by which we choose to model them, then we can avoid the kinds of paradox that have been mentioned here; and further, their avoidance points the way forward to some important principles of signal modelling.

It should be mentioned that there are other methods of modelling transient phenomena. The use of *wavelet analysis* for this, and many other purposes, is currently very poular and is of more recent origin. This theory aims to give information of the "time–scale" type rather than the Fourier "time–frequency" type; it became established in the scientific literature in the early 1980's. The theory is quite different to the Paley–Wiener theory, and we cannot enter into it here, but a chapter on wavelets will appear in H–S.

Meanwhile there are several excellent introductions to wavelet analysis, from which the Steele Prize-winning book of Daubechies (1992) can be singled out for its emphasis on the connections between wavelets, quantum mechanics and signal analysis. For the beginner the articles by Meyer (1993) and Strang (1993)[2] are very readable and informative.

17.2 Energy concentration

The usual uncertainty principle of Fourier analysis is not formulated very conveniently for a discussion of signal modelling. A different and more fruitful kind of principle is discussed in §17.3. However, we shall take a brief look at the ordinary version now, in order to contrast it with the new version to come.

The general theme of uncertainty principles in Fourier analysis is to the effect that a function cannot be very "localized" if its Fourier transform is also very localized, and vice versa. If we take "localization" to mean a small second moment then we obtain a Fourier version of the Heisenberg uncertainty principle. Actually it is usual to normalise the function and look at the second moment of its square. Thus, for a non-null function f let us put

$$\sigma_f^2 := \int_{\mathbb{R}} |tf(t)|^2 \, dt / \|f\|^2 ,$$

where the norm is that of $L^2(\mathbb{R})$, and call σ_f^2 the *spread* of f.

For an arbitrary function f the spread could be infinitely large of course.

The general principle is, then, that the spread of a function and of its Fourier transform cannot both be small. The matter can be approached by considering Gaussians. If we take $f(t) = e^{-at^2}$, $a > 0$, then as listed in Appendix A, no. 24, f is its own Fourier transform up to scale factors; in fact $f^\wedge(x) = e^{-x^2/4a}/\sqrt{2a}$. Using the special integrals

---

[2]Both conveniently located in the same issue of the Bulletin of the American Mathematical Society. Complete citations appear in the Biliography.

$$\int_{\mathbb{R}} e^{-av^2}\, dv = \sqrt{\frac{\pi}{a}} \qquad \text{and} \qquad \int_{\mathbb{R}} v^2 e^{-av^2}\, dv = \frac{1}{2a^3}\sqrt{\frac{\pi}{a}}$$

it is an easy matter to calculate $\sigma_f^2 = 1/(16a^3)$ and $\sigma_{f^\wedge}^2 = 16a^3/4$, so that $\sigma_f^2 \sigma_{f^\wedge}^2 = \frac{1}{4}$ when f is a Gaussian. It turns out that we can do no better; every function other than this Gaussian gives a larger product of spreads.

Theorem 17.1 *Let f be not null, have a continuous first derivative on $\mathbb{R}$ and be such that $tf(t) \to 0$ as $t \to \pm\infty$. Then*

$$\sigma_f^2 \sigma_{f^\wedge}^2 \geq \tfrac{1}{4}.$$

There is no extremal for this inequality other than

$$f(t) = e^{-at^2}, \qquad a > 0.$$

Proof The assertion is obvious if either variance is infinitely large. Otherwise, we shall need the inequality

$$2|\overline{f} f'| \geq \overline{f} f' + f\overline{f}' = (f\overline{f})' = |f^2|'.$$

Using Plancherel's theorem, the operational property in Appendix A no. 6, Schwarz' inequality, the inequality above and finally an integration by parts, we find that

$$\|f\|^4\, \sigma_f^2 \sigma_{f^\wedge}^2 = \int_{\mathbb{R}} |tf(t)|^2\, dt \int_{\mathbb{R}} |xf^\wedge(x)|^2\, dx = \int_{\mathbb{R}} |tf(t)|^2\, dt \int_{\mathbb{R}} |f'(x)|^2\, dx$$

$$\geq \left\{ \int_{\mathbb{R}} |t\overline{f(t)} f'(t)|\, dt \right\}^2 \geq \left\{ \tfrac{1}{2} \int_{\mathbb{R}} t|f^2(t)|'\, dt \right\}^2 = \tfrac{1}{4} \|f\|^4.$$

The reader may care to check the consequences of equality throughout the proof to find the Gaussian extremal. The result can be extended to larger classes of functions, but it is not our purpose to linger over the proof (more details can be found in Dym and McKean 1972, p. 119).

*   *   *

We seek a different measure of localization for a function, and it turns out that *energy concentration* is the key idea.

Definition 17.2 *Let I be the interval $[-\pi T, \pi T]$. The energy concentration α_f over an interval I for a function $f \in PW_B$ is defined in a natural way by*

$$\alpha_f = \frac{\int_I |f(t)|^2\, dt}{\|f\|^2}. \tag{17.2.1}$$

Our object here is to ask how large this energy concentration can be for fixed I as f ranges over PW_B. First we shall put (17.2.1) into a form which makes this maximization problem more amenable. We can write

$$\overline{f(t)} = \frac{1}{\sqrt{2\pi}} \int_B \overline{f^\wedge(y)} e^{-iyt}\, dy,$$

so the numerator in (17.2.1) can be written

$$\frac{1}{\sqrt{2\pi}} \int_I f(t) \left\{ \int_\mathbb{R} \chi_B(y)\overline{f^\wedge(y)} e^{-iyt}\, dy \right\} dt.$$

Now, by Plancherel's theorem, the expression in braces can be written

$$\int_\mathbb{R} \left(\chi_B(\cdot)e^{-i\cdot t}\right)^\vee (z)\overline{f(z)}\, dz = \int_\mathbb{R} \chi_B(z-t)\overline{f(z)}\, dz,$$

so that (17.2.1) becomes

$$\alpha_f = \frac{1}{\|f\|^2 \sqrt{2\pi}} \int_I \int_\mathbb{R} \chi_B^\vee(z-t) f(t)\overline{f(z)}\, dz\, dt.$$

However, since values of f on the complement of I do not affect the numerator of (17.2.1), we can equivalently ask for the maximum of the "integral quadratic form"

$$\frac{1}{\sqrt{2\pi}} \int_{I\times I} \chi_B^\vee(z-t)\varphi(t)\overline{\varphi(z)}\, dz\, dt$$

as φ ranges over the unit sphere of PW_B.

The answer to our maximization problem comes from the classical theory of integral equations with symmetric kernel (Courant and Hilbert 1953, p. 122 ff.). This theory shows that the maximum possible α_f is given by the largest eigenvalue of the eigenvalue equation

$$\frac{1}{\sqrt{2\pi}} \int_I \chi_B^\vee(z-t)\varphi(t)\, dt = \rho\varphi(z), \qquad (17.2.2)$$

and is attained when f is the corresponding eigenfunction. Another equally important way of solving this maximization problem is to write α_f in terms of $f^\wedge$ rather than f, as follows. From (17.2.1),

$$\alpha_f = \frac{1}{2\pi \|f^\wedge\|^2} \int_I \left\{ \int_B f^\wedge(x)e^{ixt}\, dx \right\} \left\{ \int_B \overline{f^\wedge(v)} e^{-ivt}\, dv \right\} dt$$

$$= \frac{T}{\|f^\wedge\|^2} \int_{B\times B} \operatorname{sinc} T(x-v) f^\wedge(x)\overline{f^\wedge(v)}\, dx\, dv,$$

The same principle that led us to (17.2.2) applies,[3] and we find that the maximization problem is also solved by the largest eigenvalue of the eigenvalue equation

$$T \int_B \operatorname{sinc} T(x - v)\psi(v)\, dv = \rho\psi(x), \qquad (17.2.3)$$

where ψ has been used to denote $\varphi^\wedge$. Note that $\operatorname{supp}\psi = B$. In fact, we shall find in Theorem 17.5, part 2., that (17.2.2) and (17.2.3) are effectively equivalent for our purposes in that they have the same eigenvalues.

In order to make use of the solutions we have discovered to our maximization problem we shall need much more information about these eigenvalues and eigenfunctions. Theorem 17.5 below collects some essential facts.

First we define two "restriction" operators on $L^2(\mathbb{R})$ as follows.

Definition 17.3 *Let $S \subset \mathbb{R}$. Then*

$$(T_S f)(x) := \begin{cases} f(x), & x \in S \\ 0, & x \notin S. \end{cases}$$

In particular, T_I will always mean restriction to the interval $[-\pi T, \pi T]$, an interval that we have agreed to call I. T_I is often called the operator of time-limitation. We also have the operator of band-limitation:

$$\mathcal{B}_B f := \mathcal{F}^{-1} T_B \mathcal{F} f.$$

Let us recall that a projection operator $\mathcal{P}$ on a Hilbert space has the four properties of being bounded, linear, idempotent (that is, $\mathcal{P}^2 = \mathcal{P}$) and self-adjoint.

Lemma 17.4 *The restriction operator T_S is a projection of $L^2(\mathbb{R})$ onto $L^2(S)$, and $\mathcal{B}_B$ is a projection of $L^2(\mathbb{R})$ onto PW_B.*

Proof The range of T_S is a subspace of $L^2(\mathbb{R})$ that is isometrically isomorphic to $L^2(S)$; however, by a slight abuse of notation we shall continue to call it $L^2(S)$.

That T_S is self-adjoint follows from

$$\langle T_S f, g \rangle = \int_\mathbb{R} \chi_S(t) f(t) \overline{g(t)}\, dt = \langle f, T_S g \rangle.$$

Using Plancherel's theorem we find that $\mathcal{B}_B$ is self-adjoint:

---

[3]Care must be taken when applying the standard theory of integral operators since many of the reults require the kernel to be continuous over a square domain. However, the theory can generally be extended to kernels that are piecewise continuous, by standard methods (Courant and Hilbert 1953, p. 152). This meets our requirements for domains of the type $B \times B$.

$$\langle \mathcal{B}_B f, g \rangle = \langle \mathcal{F}^{-1} T_B \mathcal{F} f, g \rangle$$
$$= \langle T_B \mathcal{F} f, \mathcal{F} g \rangle = \langle \mathcal{F} f, T_B \mathcal{F} g \rangle = \langle f, \mathcal{F}^{-1} T_B \mathcal{F} g \rangle = \langle f, \mathcal{B}_B g \rangle.$$

The other three properties for T_S and $\mathcal{B}_B$ are left as exercises for the reader.

Theorem 17.5
1. *The kernel of the integral operator on the left of (17.2.2) is of the Hilbert–Schmidt type in that it is Hermitean symmetric and belongs to $L^2(I \times I)$. As a consequence the integral operator has pure point spectrum (as defined in Definition 15.2). The eigenvalues $\{\rho_n\}$, $n \in \mathbb{N}_0$, are real, positive and decrease to 0 as n tends to ∞. Furthermore, they are simple. The corresponding eigenfunctions form an orthogonal basis for $L^2(I)$ (these eigenfunctions are called prolate spheroidal wave functions).*
2. *The operator on the left of (17.2.3) is also of the Hilbert–Schmidt type, and has the same spectrum as that on the left of (17.2.2); that is, it has eigenvalues $\{\rho_n\}$.*

The remaining parts of the theorem are only needed for the case $B = [-\pi w, \pi w]$.

3. *The eigenfunctions, $\{\psi_n\}$ say, of the operator on the left of (17.2.3) form an orthogonal basis for $L^2(B)$.*
4. *The eigenfunctions in part 1, now considered to be defined on all of $\mathbb{R}$, normalized in $L^2(\mathbb{R})$ and denoted by $\{\varphi_n\}$, form an orthonormal basis for PW_B.*
5. *The appropriate normalization for the orthogonal basis in part 1 is to write it in the form $\{\rho_n^{-1/2} T_I \varphi_n\}$.*
6. *When $f \in PW_{\pi w}$ has the expansion*

$$f = \sum_{n=0}^{\infty} a_n \varphi_n \tag{17.2.4}$$

in norm, then $T_I f$ has the expansion

$$T_I f = \sum_{n=0}^{\infty} \rho_n^{1/2} a_n \left(\rho_n^{-1/2} T_I \varphi_n \right). \tag{17.2.5}$$

in the norm of $L^2(I)$.

Proof
Part 1. The Hilbert–Schmidt properties are simple verifications. All the properties of the eigenvalues and eigenfunctions follow from the Hilbert–Schmidt theory (e.g. Riesz and Nagy 1990, pp. 230–234 and 242) except the simple property of the eigenvalues; this can be proved by a separate argument (Slepian and Pollak 1961, pp. 59–61).
Part 2. This is proved by showing that each of (17.2.3) and (17.2.2) can be transformed into the other without changing the eigenvalues. To transform (17.2.3)

into (17.2.2) let us multiply it by $e^{ixy}/\sqrt{2\pi}$ and integrate over B. On the right we obtain

$$\rho \frac{1}{\sqrt{2\pi}} \int_B e^{ixy} \psi(x)\, dx = \rho\varphi(y).$$

On the left we obtain

$$\frac{T}{\sqrt{2\pi}} \int_B e^{ixy} \int_B \operatorname{sinc} T(x-v)\psi(v)\, dv\, dx$$

$$= \frac{T}{\sqrt{2\pi}} \int_B e^{ixy} \int_B \int_I \frac{e^{-it(x-v)}}{2\pi T}\, dt\, \psi(v)\, dv\, dx$$

$$= \frac{1}{\sqrt{2\pi}} \int_I \left\{ \frac{1}{\sqrt{2\pi}} \int_{\mathbb{R}} \chi_B(x) e^{ix(y-t)}\, dx \right\} \left\{ \frac{1}{\sqrt{2\pi}} \int_B e^{itv}\psi(v)\, dv \right\}\, dt$$

$$= \frac{1}{\sqrt{2\pi}} \int_I \chi_B^{\vee}(y-t)\varphi(t)\, dt,$$

as required. In a similar way we can transform (17.2.2) into (17.2.3) by multiplying it by $e^{-ixy}/\sqrt{2\pi}$ and integrating over R.

Part 3. This follows from the Hilbert–Schmidt theory as in Part 1. Here the kernel is defined on a square containing $B \times B$, taking the value zero on the complement of $B \times B$. As such, it is piecewise continuous.

Part 4. For each $n \in \mathbb{N}_0$, φ_n is the inverse Fourier transform of ψ_n up to a constant multiple. Part 3 and the standard Fourier duality now give the required basis property.

Part 5. From Definition 17.3 and Plancherel's Theorem we can write (17.2.2) in the form

$$\mathcal{B}_B \mathcal{T}_I \varphi = \rho\varphi.$$

But φ can be replaced tautologically with $\mathcal{B}_B \varphi$ since $\mathcal{B}_B$ is a projection onto PW_B. After this change has been made we can write (17.2.2) in the form

$$\mathcal{B}_B \mathcal{T}_I \mathcal{B}_B \varphi = \rho\varphi.$$

Now using the fact that $\mathcal{B}_B$ and $\mathcal{T}_I$ are self-adjoint, the required normalization is obtained as follows:

$$\langle \mathcal{T}_I \varphi_i, \mathcal{T}_I \varphi_j \rangle = \langle \mathcal{T}_I \mathcal{B}_B \varphi_i, \mathcal{T}_I \mathcal{B}_B \varphi_j \rangle = \langle \mathcal{B}_B \mathcal{T}_I^2 \mathcal{B}_B \varphi_i, \varphi_j \rangle$$
$$= \langle \mathcal{B}_B \mathcal{T}_I \mathcal{B}_B \varphi_i, \varphi_j \rangle = \langle \rho_i \varphi_i, \varphi_j \rangle = \rho_i \delta_{ij}.$$

Part 6. The expansion (17.2.5) follows at once from (17.2.4) since $\mathcal{T}_I$ is a bounded linear operator.

17.3 The uncertainty principle of signal theory

We now make final preparations for stating and proving the uncertainty principle. In this section we shall take B to be a single interval.

Let f, $g \in L^2(\mathbb{R})$. Then by Schwarz' inequality $|\mathfrak{Re}\langle f, g \rangle| \leq \|f\| \|g\|$, so that when f and g are not null we have

$$\frac{|\mathfrak{Re}\langle f, g \rangle|}{\|f\| \|g\|} \leq 1,$$

and the following definition is a natural consequence.

Definition 17.6 *The angle between f and g is defined by*

$$\theta(f, g) = \arccos\left\{ \frac{\mathfrak{Re}\langle f, g \rangle}{\|f\| \|g\|} \right\}.$$

Two subspaces S_1 and S_2 of $L^2(\mathbb{R})$ are said to be separated by a least angle if $\inf_{f \in S_1, g \in S_2} \theta(f, g) > 0$ is attained by some non-null $f_0 \in S_1$ and $g_0 \in S_2$.

Lemma 17.7 *Let $f \in PW_B$ be given. Then*

$$\inf_{g \in L^2(I)} \theta(f, g) = \arccos\left\{ \frac{\|T_I f\|}{\|f\|} \right\} > 0.$$

Proof We have

$$\mathfrak{Re}\langle f, g \rangle = \mathfrak{Re}\langle f, T_I g \rangle = \mathfrak{Re}\langle T_I f, g \rangle \leq |\langle T_I f, g \rangle| \leq \|T_I f\| \|g\|,$$

equality holding in both inequalities if $T_I f$ is a positive constant multiple of g. Hence

$$\frac{\mathfrak{Re}\langle f, g \rangle}{\|f\| \|g\|} \leq \frac{\|T_I f\|}{\|f\|}.$$

Because the principal branch of arccos is monotonic decreasing, we have

$$\theta(f, g) \geq \arccos\left\{ \frac{\|T_I f\|}{\|f\|} \right\}$$

When equality occurs, the infimum of θ is attained; but we have just seen that equality can indeed occur and this gives the required result.

The following theorem is effectively an uncertainty principle in signal analysis. It says that if the parameters T and w are given, then a member of $L^2(I)$ and a member of $PW_{\pi w}$ must be separated in $L^2(\mathbb{R})$ by at least a certain minimum angle.

Theorem 17.8 *Let $B = [-\pi w, \pi w]$. Let ρ_0 be the largest eigenvalue of (17.2.2) and φ_0 the corresponding eigenfunction. Then $PW_{\pi w}$ and $L^2(I)$ are separated by the least angle $\arccos \sqrt{\rho_0}$, and this least angle is attained by $\varphi_0 \in PW_{\pi w}$ and $T_I \varphi_0 \in L^2(I)$.*

Proof By the previous lemma,

$$\inf_{f \in B} \inf_{g \in L^2(I)} \theta(f, g) = \inf_{f \in B} \arccos \left\{ \frac{\|T_I f\|}{\|f\|} \right\}. \tag{17.3.1}$$

By Theorem 17.5, part 6, we can now expand f and $T_I f$ in the series (17.2.4) and (17.2.5) respectively. From Theorem 17.5, parts 4 and 5, the corresponding Parseval relations are

$$\|f\|^2 = \sum_{k=0}^{\infty} |a_k|^2 \quad \text{and} \quad \|T_I f\|^2 = \sum_{k=0}^{\infty} \rho_k |a_k|^2.$$

These allow us to write

$$\frac{\|T_I f\|^2}{\|f\|^2} = \frac{\sum_{k=0}^{\infty} \rho_k |a_k|^2}{\sum_{k=0}^{\infty} |a_k|^2}. \tag{17.3.2}$$

But from Theorem 17.5, part 1, $\rho_k \leq \rho_0$, $k > 0$, so we have

$$\frac{\|T_I f\|^2}{\|f\|^2} \leq \rho_0 \frac{\sum_{k=0}^{\infty} |a_k|^2}{\sum_{k=0}^{\infty} |a_k|^2} = \rho_0, \tag{17.3.3}$$

and equality is attained by taking that f for which $a_k = 0$, $k > 0$, in (17.3.2); that is, $f = a_0 \varphi_0$. This, coupled with (17.3.1) and the fact that arccos is monotonic decreasing, gives the required least angle. Furthermore, (17.3.3) shows that this least angle can indeed be attained by taking $\varphi_0 \in PW_{\pi w}$ and $T_I \varphi_0 \in L^2(I)$.

The uncertainty principle embodied in the previous theorem can be put into terms of energy concentrations of functions and their Fourier transforms, and given a geometrically appealing formulation if we allow the following rather heuristic argument.

We note first that if $f \in L^2(\mathbb{R})$ then $\alpha_f = \|T_I f\|/\|f\|$ and $\alpha_{f^\wedge} = \|B_B f\|/\|f\|$ are the energy concentrations of f on I and of $f^\wedge$ on B respectively, and they are also the cosines of the angles made by f with D and PW_B respectively. Geometric intuition suggests that the sum of these angles cannot be less than the least angle between $L^2(I)$ and PW_B, so we have

$$\arccos \alpha_f + \arccos \alpha_{f^\wedge} \geq \arccos \sqrt{\rho_0}. \tag{17.3.4}$$

This can be interpreted graphically by noting that the set of points $(\alpha_f, \alpha_{f^\wedge})$ in the unit square, which represents all the possible time and frequency concentrations available as f ranges over $L^2(\mathbb{R})$, must remain within the region common to the unit square and the region delimited by (17.3.4),[4] and this is bounded away from $(1, 1)$.

---

[4]This is the interior of an ellipse. A diagram can be found in Slepian (1983, p. 388).

It was in this form that the uncertainty principle was given by Landau and Pollak; further explanations, diagrams and references can be found in Landau (1985) and Slepian (1983). Another treatment is in Dym and McKean (1972, §2.9*).

17.4   Prolate spheroidal wave functions

Let us take B to be a single interval $[-\pi w, \pi w]$. We have seen that the eigenfunctions $\{\varphi_n\}$ of (17.2.2) have a remarkable 'double orthogonality' property, expressed in Theorem 17.5, parts 1 and 4. Furthermore, it follows from the properties of $\{\rho_k\}$ discussed in Section 17.2 that ρ_0 is the largest energy concentration attainable by a member of PW_B; and that member is φ_0. Again, ρ_1 is the largest energy concentration attainable by a member of the PW_B that is orthogonal to φ_0; and that member is φ_1. Again, ρ_2 is the largest energy concentration attainable by a member of the PW_B that is orthogonal both to φ_0 and to φ_1; and that member is φ_2—and so on.

Another property worth emphasizing is that in the uncertainty principle of the previous section, the band-limited function φ_0 and its time-limited version $T_I \varphi_0$ were found to be as close as two such functions can be.

The members φ_n of this very remarkable function set are called *prolate spheroidal wave functions*. The search by engineers for functions that would be band-limited and as nearly as possible time-limited was successfully completed by Landau, Pollak and Slepian of the (then) Bell Telephone Laboratories, in a sequence of papers dating from the mid-1960s, when they discovered, among many other important related results, that the prolate spheroidal wave functions lie at the heart of the band-limited signal modelling problem. This beautiful and subtle corpus of work, affectionately known as the 'Bell papers' and on which nearly all the present chapter is based, can be traced through the expository articles of Landau (1985) and Slepian (1976, 1983). A very readable entrée to the prolate spheroidal wave functions in this context can be found in the book of Papoulis (1977, §6-4).

17.5   The Nyquist–Landau minimal sampling rate

We now prove a theorem that has been quoted many times in the previous pages. As a final preparatory remark, we note that the energy concentration defined in (17.2.1) shows that the eigenvalues $\{\rho_k\}$ lie between 0 and 1; in fact, none can equal 1 because no Paley–Wiener function can have *all* its energy concentrated on I.

Theorem 17.9 *Let B be the union of finitely many real intervals, and let a typical one be denoted by β and have length $2b$. Let $\{\lambda_n\}$, $n \in \mathbb{Z}$, be a set of stable sampling for PW_B. Then the density (Definition 10.3) $D(\lambda_n)$ of this set satisfies*

$$D(\lambda_n) \geq \frac{m(B)}{2\pi}. \tag{17.5.1}$$

Before giving the details, an intuitive guide to the proof is in order. From PW_B we select that sub-class WC whose members have their energy well-concentrated on I. Let s be the number of sample points λ_n occurring in I. Because of the stability property we can set to zero the samples of $f \in WC$ taken from sample points outside I without substantially affecting the reconstruction procedure. Then any member of WC is effectively determined by s measurements, its values at s sample points. The dimension of WC is approximately s, because there are s linearly independent members of WC that vanish at all s sample points except one; and any member of WC is a linear combination of these. Now, if we recall the remarks made in the previous section about the energy concentration properties of φ_0, φ_1, ... , etc., we find that, because the eigenvalues ρ_n are simple, the number of linearly independent members of WC is equal to the number of the ρ_n that are closest to 1; that is, those giving the largest energy concentrations. If we can estimate this number, then we can estimate s. Then it will be a simple matter to calculate the density by letting I become large.

In the proof, C's with subscripts denote various constants.

Proof *of Theorem 17.9.* We proceed to estimate the number of "large" eigenvalues, that is; those close to 1.

We shall need two formulae from the theory of integral operators (they can be found in Riesz and Nagy 1990, p. 243). The first is the *trace* of the operator in (17.2.3), given by

$$\sum_{k=0}^{\infty} \rho_k = T \int_B [\mathrm{sinc}(x - y)]_{y=x}\, dx = Tm(B) = m(B)m(I)/2\pi. \quad (17.5.2)$$

The second is the *Hilbert–Schmidt norm*, given by

$$\sum_{k=0}^{\infty} \rho_k^2 = T^2 \int_{B\times B} \mathrm{sinc}^2\, T(v - t)\, dv\, dt \quad (17.5.3)$$

$$\geq T^2 \sum_{\beta} \int_{\beta\times\beta} \mathrm{sinc}^2\, T(v - t)\, dv\, dt, \quad (17.5.4)$$

The last summation is, of course, over all components β of B. On making a sequence of self-evident changes of variable, we can pass from the summand of (17.5.4) to

$$T^2 \int_{-b}^{b} \int_{-b}^{b} \mathrm{sinc}^2\, T(s - t)\, ds\, dt = (Tb)^2 \int_{-1}^{1} \left\{ \int_{-1-\tau}^{1-\tau} \mathrm{sinc}^2\, Tbv\, dv \right\} d\tau$$

$$= (Tb)^2 \int_{-1}^{1} \Phi(\tau)\, d\tau, \quad (17.5.5)$$

say, where we have put

$$\Phi(\tau) = \int_{-1-\tau}^{1-\tau} \operatorname{sinc}^2 Tbv \, dv,$$

and consequently we have, by the standard calculus rule,

$$\Phi'(\tau) = \operatorname{sinc}^2 Tb(1+\tau) - \operatorname{sinc}^2 Tb(1-\tau).$$

These enable us to integrate by parts in (17.5.5), whereupon it becomes

$$4(Tb)^2 \int_0^2 \operatorname{sinc}^2 Tbv \, dv - 2(Tb)^2 \int_0^2 y \operatorname{sinc}^2 Tby \, dy. \qquad (17.5.6)$$

Let us add and subtract to (17.5.6) the integral

$$4(Tb)^2 \int_2^\infty \operatorname{sinc}^2 Tbv \, dv,$$

and use the special integral

$$\int_0^\infty \operatorname{sinc}^2 av \, dv = \frac{1}{2a}.$$

Then we can write (17.5.6) in the form

$$2bT - 4(Tb)^2 \int_2^\infty \operatorname{sinc}^2 Tbv \, dv - 2(Tb)^2 \int_0^2 y \operatorname{sinc}^2 Tby \, dy. \qquad (17.5.7)$$

In modulus the middle term of (17.5.7) is not greater than

$$\frac{4}{\pi^2} \int_2^\infty \frac{1}{v^2} \, dv = \frac{2}{\pi^2}.$$

For $T > 1/2$ we can write the last term of (17.5.7) in the form

$$-\frac{2}{\pi^2} \left\{ \int_0^{1/T} \frac{\sin^2 \pi Tbw}{w} \, dw + \frac{1}{2} \int_{1/T}^2 \frac{1 - \cos 2\pi Tbw}{w} \, dw \right\}$$

$$\geq -\frac{2}{\pi^2} \left\{ (\pi Tb)^2 \int_0^{1/T} w \, dw + \int_{1/T}^2 \frac{1}{w} \, dw \right\}$$

$$= -2(Tb)^2 \frac{1}{2T^2} - \frac{2}{\pi^2} \left(\log 2 - \log \frac{1}{T} \right) = C_1 - C_2 \log T.$$

These estimates applied to (17.5.7) together with (17.5.5) and (17.5.4) show that

$$\sum_{k=0}^{\infty} \rho_k^2 \geq \sum_{\beta} \{2bT - C_3 \log T + C_4\}$$

$$= \frac{m(B)m(I)}{2\pi} - C_5 \log T + C_6.$$

This estimate combined with (17.5.2) shows that

$$\sum_{k=0}^{\infty} \rho_k(1 - \rho_k) \leq C_7 \log T + C_8. \qquad (17.5.8)$$

Those ρ_k which are close to 0 and those close to 1 contribute little to this sum. Consider therefore those "intermediate" ρ_k for which $0 < \delta \leq \rho_k \leq \gamma < 1$, where δ and γ are fixed. Each contributes to (17.5.8) an amount not less than

$$a = \min[\delta(1 - \delta), \gamma(1 - \gamma)].$$

This means that the number of these intermediate eigenvalues cannot exceed $(C_7 \log T + C_8)/a$; that is, it can grow at most logarithmically with T. Therefore from (17.5.2) the number of "large" eigenvalues, those near 1, must behave asymptitcally like $m(B)m(I)/2\pi$. However, in the remarks we made just before embarking on this proof, we saw that this number is an estimate for the number of sample points occurring in the interval $I = [-\pi T, \pi T]$. If we now divide this number by $m(I)$ and let $T \to \infty$, the result is to calculate the density of the sample points over $\mathbb{R}$. Finally then, the density of $\{\lambda_k\}$ over $\mathbb{R}$ is not less than $m(B)/2\pi$.

APPENDIX A

FOURIER TRANSFORMS

Let $\mathcal{F}$ denote the Fourier transform, and let $\mathcal{F}_c$ and $\mathcal{F}_s$ denote the Fourier cosine transform and the Fourier sine transform respectively, where

$$f^{\wedge}(x) = (\mathcal{F}f)(x) = \frac{1}{\sqrt{2\pi}} \int_{\mathbf{R}} f(t)e^{-ixt}\, dt,$$

$$(\mathcal{F}_c f)(x) = \sqrt{\frac{2}{\pi}} \int_0^{\infty} f(t) \cos xt\, dt, \qquad (\mathcal{F}_s f)(x) = \sqrt{\frac{2}{\pi}} \int_0^{\infty} f(t) \sin xt\, dt.$$

Some useful formulae are

$$\mathcal{F}^{-1} \equiv \mathcal{F} \equiv \mathcal{F}_c, \quad \text{when } f \text{ is even.}$$
$$\mathcal{F}^{-1} \equiv -\mathcal{F} \equiv -i\,\mathcal{F}_s, \quad \text{when } f \text{ is odd.}$$
$$\left(\mathcal{F}^2 f\right)(t) = f(-t).$$

Some operational formulae

Function	Fourier transform
1. $f(t)\cos At,\ A > 0$	$\frac{1}{2}\left\{f^\wedge(x+A)+f^\wedge(x-A)\right\}$
2. $f(t)\sin At,\ A > 0$	$\frac{i}{2}\left\{f^\wedge(x+A)-f^\wedge(x-A)\right\}$
3a. $f(t+h),\ h \in \mathbb{R}$	$e^{ihx}\, f^\wedge(x)$
3b. $e^{ikt}f(t),\ k \in \mathbb{R}$	$f^\wedge(x-k)$
4. $f(rt),\ r \in \mathbb{R}$	$\dfrac{1}{\lvert r\rvert}f^\wedge\left(\dfrac{x}{r}\right)$
5. $f^\sim(t)$	$-i\,\operatorname{sgn} x\, f^\wedge(x)$
6. $(-i\mathcal{D})^n\, f(t)$	$x^n\, f^\wedge(x)$
7. $(f * g)(t)$	$f^\wedge(x)\, g^\wedge(x)$

In these operational formulae, $f^\sim$ denotes the Hilbert transform of f (see Appendix B), $\mathcal{D}^n$ denotes n-fold differentiation and $f*g$ denotes the convolution $(2\pi)^{-1/2}\int_{\mathbb{R}} f(v)g(t-v)\,dv$.

Some special Fourier transforms

With the exception of the last two entries, the table to follow is a collection of special band-limited functions, most of which are needed in the text. They have been gathered here for utility and interest. Their band-regions are shown at the extreme right, so that the transforms in the right-hand column are null outside the stated range of x.

Item 15 is a very striking formula of Ramanujan (see, e.g., Titchmarsh 1948, p. 187). The function f in item 8 is sometmes called a *kernel of the Dirchlet type*, and that in item 9 a *kernel of the Fejér type*. The function f in item 21 is called the *kernel of Bohman–Zheng Wei-xing*; the function f in item 22 is called the *kernel of the de la Valleé Poussin means*; and the function f in item 23 is called the *kernel of Cauchy–Poisson*. Another kernel can be obtained as a special case of item 20, by dividing both sides by b^ν and taking the limit as $b \to 0$. Then the function f becomes $J_{\nu+1/2}(|\pi wt|)/|\pi wt|^{\nu+1/2}$, the *kernel of Bochner–Riesz*, with Fourier transform $[2^\nu\Gamma(\nu+1)]^{-1}\{1-(x/\pi w)^2\}^\nu\chi_{\pi w}(x)$. The special case $\nu = -\frac{1}{2}$ gives a simple oscillatory function with an unbounded Fourier transform.

All the kernels mentioned here have been discussed in the context of generalized sampling series by Splettstößer (1978).

A rich source of information on special functions is the book of Magnus *et al.* (1966).

$f(t)$	$f^{\wedge}(x) = \dfrac{1}{\sqrt{2\pi}} \displaystyle\int_{\mathbb{R}} f(t)e^{-ixt}\, dt$
8. $\operatorname{sinc} wt$	$\dfrac{1}{w\sqrt{2\pi}} \qquad\qquad \lvert x\rvert < \pi w,$ $\dfrac{1}{2w\sqrt{2\pi}} \qquad\qquad \lvert x\rvert = \pi w.$
9. $\operatorname{sinc}^2 \tfrac{1}{2}wt$	$\dfrac{2}{w\sqrt{2\pi}}\left(1 - \dfrac{\lvert x\rvert}{\pi w}\right) \qquad \lvert x\rvert \le \pi w.$
10. $\operatorname{sinc}^3 \tfrac{1}{3}wt$	$\dfrac{9}{4w\sqrt{2\pi}}\left[1 - 3\left(\dfrac{x}{\pi w}\right)^2\right] \qquad \lvert x\rvert \le \dfrac{\pi w}{3},$ $\dfrac{27}{8w\sqrt{2\pi}}\left(1 - \dfrac{\lvert x\rvert}{\pi w}\right)^2 \qquad \dfrac{\pi w}{3} \le \lvert x\rvert \le \pi w.$
11. $\operatorname{sinc} \tfrac{1}{2}wt \sin \tfrac{1}{2}\pi wt$	$\dfrac{-i}{w\sqrt{2\pi}}\, \operatorname{sgn} x \qquad\qquad \lvert x\rvert < \pi w.$
12. $(1 + i)\operatorname{sinc} \tfrac{1}{2}wt$ $\times \left(\cos \tfrac{1}{2}\pi wt + \sin \tfrac{1}{2}\pi wt\right)$	$\dfrac{i}{w}\sqrt{\dfrac{2}{\pi}} \qquad\qquad -\pi w < x \le 0,$ $\dfrac{1}{w}\sqrt{\dfrac{2}{\pi}} \qquad\qquad 0 < x \le \pi w.$

$f(t)$	$f^\wedge(x) = \dfrac{1}{\sqrt{2\pi}} \displaystyle\int_{\mathbb{R}} f(t)e^{-ixt}\, dt$	
13. $\dfrac{1 - \operatorname{sinc} t}{t}$	$\dfrac{i}{\sqrt{2\pi}}\,(x + \pi)$	$-\pi \le x < 0,$
	$\dfrac{i}{\sqrt{2\pi}}\,(x - \pi)$	$0 < x \le \pi.$
14. $\dfrac{1 - \operatorname{sinc} t}{t^2}$	$\dfrac{1}{2\sqrt{2\pi}}\,(\pi - |x|)^2$	$|x| \le \pi.$
15. $\dfrac{1}{\Gamma(\alpha + t)\,\Gamma(\beta - t)}$ $\alpha + \beta > 1$	$\dfrac{e^{ix(\alpha-\beta)/2}\left(2\cos\frac{1}{2}x\right)^{\alpha+\beta-2}}{\sqrt{2\pi}\,\Gamma(\alpha + \beta - 1)}$	$|x| < \pi.$
16. $\operatorname{sinc} wt \,\cos ct$ $c > \pi w$	$\dfrac{1}{2w\sqrt{2\pi}}$	$c - \pi w < |x| < c + \pi w.$
17. $\operatorname{sinc} wt \,\sin ct$ $c > \pi w$	$\dfrac{1}{2w\sqrt{2\pi}}(-i\operatorname{sgn} x)$	$c - \pi w < |x| < c + \pi w.$
18. $\operatorname{sinc}\frac{1}{2}t \sin^2\frac{1}{4}\pi t$	$\dfrac{1}{2\sqrt{2\pi}}$	$|x| < \dfrac{\pi}{2},$
	$-\dfrac{1}{2\sqrt{2\pi}}$	$\dfrac{\pi}{2} < |x| < \pi.$

$f(t)$	$f^\wedge(x) = \dfrac{1}{\sqrt{2\pi}} \displaystyle\int_{\mathbb{R}} f(t)e^{-ixt}\, dt$	
19. $i^n t^{-1/2} J_{n+1/2}(t)$ $n \in \mathbb{N}_0$	$P_n(x)$	$\lvert x \rvert \leq 1.$
20. $ab^\nu \dfrac{J_{\nu+1/2}(\sqrt{b^2 + a^2 t^2})}{(b^2 + a^2 t^2)^{(\nu+1/2)/2}}$ $a > 0,\ b > 0,\ \nu > -1$	$\left[1 - \left(\dfrac{x}{a}\right)^2\right]^{\nu/2} J_\nu\left(b\sqrt{1 - \left(\dfrac{x}{a}\right)^2}\right)$	$\lvert x \rvert \leq a.$
21. $\left(\dfrac{\cos \pi w t/2}{1 - (wt)^2}\right)^2$	$\dfrac{1}{2w}\sqrt{\dfrac{\pi}{2}}\left\{\pi\left(1 - \dfrac{\lvert x \rvert}{\pi w}\right)\cos(x/w)\right.$ $\left. + \sin(\lvert x \rvert/w)\right\}$	$\lvert x \rvert \leq \pi w.$
22. $\operatorname{sinc}\tfrac{3}{4}wt \ \operatorname{sinc}\tfrac{1}{4}wt$	$\dfrac{2}{3w}\sqrt{\dfrac{2}{\pi}}$ $\dfrac{4}{3w}\sqrt{\dfrac{2}{\pi}}\left(1 - \dfrac{\lvert x \rvert}{\pi w}\right)$	$\lvert x \rvert \leq \dfrac{\pi w}{2},$ $\dfrac{\pi w}{2} \leq \lvert x \rvert \leq \pi w.$
23. $\dfrac{1}{1 + t^2}$	$\sqrt{\dfrac{\pi}{2}}e^{-\lvert x \rvert}$	$x \in \mathbb{R}.$
24. e^{-at^2}	$\dfrac{1}{\sqrt{2a}}e^{-x^2/4a}$	$x \in \mathbb{R}.$

APPENDIX B

HILBERT TRANSFORMS

The Hilbert transform $\mathcal{H}f$ of a function f is defined formally by

$$(\mathcal{H}f)(x) = f^{\sim}(x) = \lim_{\epsilon \to 0} \frac{1}{\pi} \int_{|t| \geq \epsilon > 0} \frac{f(x-t)}{t} \, dt.$$

It is an example of a singular convolution operator, and belongs to the class of Calderón–Zygmund operators. An excellent introduction to the Hilbert transform can be found in Stein and Weiss (1971).

The limit in the definition exists almost everywhere whenever $f \in L^p(\mathbb{R})$, $1 < p < \infty$, and then $\mathcal{H}$ is a bounded linear operator. This fact is expressed by the *M. Riesz inequality*: $\|f^{\sim}\|_p \leq A_p \|f\|_p$, in which A_p is a constant depending only on p. The Hilbert transform is a unitary operator on $L^2(\mathbb{R})$.

The Hilbert transform $\mathcal{H}$ has been called "one of the most important operators in analysis" (Beals *et al.* 1992). This is because of its central rôle in understanding the structure of natural operations on functions, particularly relations between differentiability properties of functions and their harmonic or holomorphic extensions.

The Hilbert transform can be given an alternative definition, for example when $f \in L^2(\mathbb{R})$, which reveals that it is a simple multiplier operator with respect to the Fourier transform. Thus

$$\mathcal{H}f = \mathcal{F}^{-1}\left(-i\,\mathrm{sgn}(\cdot)\right)\mathcal{F}f.$$

The operational property $f^{\sim\sim} = -f$ seems to have been noticed first by Hilbert, and so $\mathcal{H}$ acquired its name.

Some special Hilbert transforms

$f(t)$	$f^{\sim}(x)$
1. $\cos t$	$\sin x$
2. $\mathrm{sinc}\, at$	$\mathrm{sinc}\,\tfrac{1}{2}ax \sin \tfrac{1}{2}\pi ax$
3. $\mathrm{sinc}^2 \tfrac{1}{2} t$	$\dfrac{2}{\pi^2} \dfrac{\pi x - \sin \pi x}{x^2}$
4. $\mathrm{sinc}\, t \cos at$	$\mathrm{sinc}\, x \sin ax$

REFERENCES

1. Annaby, M.H. On Kramer's theorem associated with second order boundary-value problems. *J. Appl. Anal.* (to appear).
2. Apostol, T.M. (1957). *Mathematical analysis: a modern approach to advanced calculus.* Addison-Wesley, Reading, MA.
3. Beals, R.W., Coifman, R.R. and Jones, P.W. (1992). Commentary on Calderón's research. In: Alberto P. Calderón receives National Medal of Science. *Notices Am. Math. Soc.*, **39**, 283–285.
4. Beaty, M.G. (1994). Multichannel sampling for multiband signals. *Signal Proc.*, **36**, 133–138.
5. Beaty, M.G. and Dodson, M.M. (1989). Derivative sampling for multiband signals. *Num. Funct. Analysis Optim.*, **10**, 875–898.
6. Beaty, M.G. and Dodson, M.M. (1993). The distribution of sampling rates for signals with equally wide, equally spaced spectral bands. *SIAM J. Appl. Math.*, **53**, 893–906.
7. Beaty, M.G. and Higgins, J.R. (1994). Aliasing and Poisson summation in the sampling theory of Paley–Wiener spaces. *J. Fourier Analysis Applic.*, **1**, 67–85.
8. Beaty, M.G., Dodson, M.M. and Higgins, J.R. (1994). Approximating Paley–Wiener functions by smoothed step functions. *J. Approx. Theory*, **78**, 433–445.
9. Benedetto, J.J. (1992). Irregular sampling and frames. In: *Wavelets: a tutorial in theory and applications* (ed. C.K. Chui), 445–507, Academic Press, London.
10. Beurling, A. (1989). *Collected works of Arne Beurling. Vol. 1. Complex analysis. Vol. 2. Harmonic analysis* (eds L. Carleson, P. Malliavin, J. Neuberger, and J. Wermer). Birkhaüser-Verlag, Basel.
11. Beutler, F.J. (1966). Error-free recovery of signals from irregularly spaced samples. *SIAM Rev.*, **8**, 328–335.
12. Bezuglaya, L. and Katsnelson, V. (1993). The sampling theorem for functions with limited multi-band spectrum. *Z. Anal. Anwendungen*, **12**, 511–534.
13. Bilinskis, I. (1995). Deliberately pseudorandomized DSP — advantages and limitations. In: *1995 workshop on sampling theory and applications.* Conference Proceedings, Jurmala, Latvia, Sept. 1995, pp 201–207. Institute of Electronics and Computer Science, Riga, Latvia.
14. Bilinskis, I. and Mikelsons, A. (1992). *Randomized signal processing.* Prentice Hall, New York.
15. Boas, R.P. (1954). *Entire functions.* Academic Press, New York.
16. Boas, R.P. (1972). Summation formulas and band-limited signals. *Tôhoku Math. J.*, **24**, 121–125.
17. Boas, R.P. and Pollard, H. (1973). Continuous analogues of series. Amer. Math. Monthly, **80**, 18–25.

18. Brown, J.L. Jr. (1967). On the error in reconstructing a non-bandlimited function by means of the bandpass sampling theorem. *J. Math. Analysis Applics*, **18**, 75–84; Erratum (1968). Ibid., **21**, 699.

19. Brown, J.L. Jr. (1985). An RKHS analysis of sampling theorems for harmonic-limited signals. *IEEE Trans. Acoust. Speech Signal Proc.*, ASSP-**33**, 437–440.

20. Brown, J.L. Jr. (1990). Summation of certain series using the Shannon sampling theorem. *IEEE Trans. Education*, **33**, 337–340.

21. Brown, J.L. Jr. (1993). Sampling of bandlimited signals: fundamental results and some extensions. In: *Handbook of Statistics, Vol. 10* (eds N.K. Bose and C.R. Rao). Elsevier, Amsterdam, 59–101.

22. Burkhardt, H. (1899–1916). *Trigonometrische Interpolation.* Enzyklopädie Math. Wiss. (Band) II-7, (Heft) 5, Teubner, Leipzig.

23. Butzer, P.L. (1983). A survey of the Whittaker–Shannon sampling theorem and some of its extensions. *J. Math. Res. Expos.*, **3**, 185–212.

24. Butzer, P.L. (1994). The Hausdorff–Young theorems of Fourier analysis and their impact. *J. Fourier Anal. Appl.*, **1**, 113–130.

25. Butzer, P.L. and Nasri-Roudsari. Kramer's sampling theorem of signal analysis and its role in mathematics; a survey. In: *Image processing; mathematical methods and applications.* I.M.A. Conference Proceedings, Cranfield University, UK, Sept. 1994. Oxford University Press. (To appear.)

26. Butzer, P.L. and Nessel, R.J. (1971). *Fourier analysis and approximation, Vol. 1. One-dimensional theory.* Academic Press, New York.

27. Butzer, P.L. and Schöttler, G. (1993). Sampling expansions for Jacobi functions and their applications. *Acta Sci. Math. (Szeged)*, **57**, 305–327.

28. Butzer, P.L. and Schöttler, G. (1994). Sampling theorems associated with fourth- and higher-order self-adjoint eigenvalue problems. *J. Comput. Appl. Math.*, **51**, 159–177.

29. Butzer, P.L. and Splettstößer, W. (1977). *Approximation und interpolation durch verallgemeinerte Abtastsummen.* Westdeutscher Verlag, Opladen.

30. Butzer, P.L. and Stens, R.L. (1992). Sampling theory for not necessarily band-limited functions; a historical overview. *SIAM Rev.*, **34**, 40–53.

31. Butzer, P.L. and Stens, R.L. (1993). Linear prediction by samples of the past. Chapter 5 in: *Advanced topics in Shannon sampling and interpolation theory* (ed. R.J. Marks II). Springer-Verlag, New York.

32. Butzer, P.L. and Stens, R.L. De la Vallée Poussin's paper of 1908 on interpolation and sampling theory, and its influence. In: *Charles Baron de la Vallée Poussin, Collected Works* (eds J. Mawhin and P.L. Butzer). (To appear.)

33. Butzer, P.L., Hauss, M. and Stens, R.L. (1991). The sampling theorem and its unique role in various branches of mathematics. *Festschrift zum 300-jährigen Bestehen der Gesellschaft. Mitt. Math. Ges. Hamburg* **12**, 523–547.

34. Butzer, P.L., Splettstößer, W. and Stens, R. L. (1988). The sampling theorem and linear prediction in signal analysis, *Jahresber. Deutsch. Math-Verein.*, **90**, 1–70.

35. Campbell, L.L. (1985). Further results on a series representation related to the sampling theorem. *Signal Proc.*, **9**, 225–231.

36. Cheung, K.W. (1993). A multidimensional extension of Papoulis' generalized sampling expansion with applications in minimum density sampling. In: *Advanced topics in Shannon sampling and interpolation theory* (ed. R.J. Marks II). Springer-Verlag, New York, 85–119.

37. Coddington, E.A. and Levinson, N. (1955). *Theory of differential equations.* McGraw-Hill, New York.

38. Coifman, R.R. and Weiss, G. (1977). Extensions of Hardy spaces and their use in analysis, *Bull. Am. Math. Soc.*, **83**, 569–645.

39. Courant, R. and Hilbert, D. (1953). *Methods of mathematical physics*, Vol. I. Interscience, New York.

40. Daubechies, I. (1992). *Ten lectures on wavelets.* CBMS-NSF Regional Conference Series in Applied Mathematics, Vol. 61, SIAM, Philadelphia.

41. Dodson, M.M. (1992). Shannon's sampling theorem. *Curr. Sci.*, **63**, 253–260.

42. Dodson, M.M. and Silva, A.M. (1985). Fourier analysis and the sampling theorem. *Proc. R. Irish Acad.*, **85** A, 81–108.

43. Dodson, M.M. and Silva, A.M. (1989). An algorithm for optimal regular sampling. *Signal Proc.*, **17**, 169–174.

44. Dodson, M.M., Silva, A.M. and Soucek, V. (1986). A note on Whittaker's cardinal series in harmonic analysis. *Proc. Edinb. Math. Soc.*, **29**, 349–357.

45. Dunford, N. and Schwartz, J.T. (1988). *Linear operators Part I: general theory.* Wiley Classics Library Edition, Wiley Interscience, New York.

46. Dym, H. and McKean, H.P. (1972). *Fourier series and integrals.* Academic Press, New York.

47. Everitt, W.N., Schöttler, G. and Butzer, P.L. (1994). Sturm–Liouville boundary value problems and Lagrange interpolation series. *Rend. Mat. Appl.*, **14**, 87–126.

48. Feichtinger, H.G. and Gröchenig, K. (1992). Iterative reconstruction of multivariate band-limited functions from irregular sampling values. *SIAM J. Math Anal.*, **23**, 244–261.

49. Freeden, W. (1984). Spherical spline interpolation — basic theory and computational aspects. *J. Comput. Appl. Math.*, **11**, 367–375.

50. Goldman, S. (1953). *Information theory.* Prentice-Hall, New York.

51. Grechikhin, A.E. (1969). The possibility of simplifying the analysis of band signals, representing the sum of discrete readings (Russian). *Radiotekhnika*, **24**, 104–106.

52. Griffith, J.L. (1955). Hankel transforms of functions zero outside a finite interval. *J. Proc. R. Soc. NSW*, **89**, 109–115.

53. Hardy, G.H. (1939). Notes on special systems of orthogonal functions, II: On functions orthogonal with respect to their own zeros. *J. Lond. Math. Soc.*, **14**, 37–44.

54. Hardy, G.H. (1941). Notes on special systems of orthogonal functions, IV: The orthogonal functions of Whittaker's cardinal series. *Proc. Camb. Phil. Soc.*, **37**, 331–348.

55. Hardy, G.H., Littlewood, J.E. and Pólya, G. (1934). *Inequalities.* Cambridge University Press, London.

56. Higgins, J.R. (1972). An interpolation series associated with the Bessel–Hankel transform. *J. Lond. Math. Soc. (2)*, **5**, 707–714.

57. Higgins, J.R. (1977). *Completeness and basis properties of sets of special functions*. Cambridge University Press, London.

58. Higgins, J.R. (1985). Five short stories about the cardinal series. *Bull. Am. Math. Soc.*, **12**, 45–89.

59. Higgins, J.R. (1987a). Finite spherical analogues of the Whittaker–Shannon sampling theorem. In: *6. Aachener Symposium für Signaltheorie. Mehrdimensionale Signale und Bildverarbeitung*, Proceedings, Springer-Verlag, Berlin, 49–51.

60. Higgins, J.R. (1987b). Some gap sampling series for multiband signals. *Signal Proc.*, **12**, 313–319.

61. Higgins, J.R. (1991). Sampling theorems and the contour integral method. *Applic. Anal.*, **41**, 155–171.

62. Higgins, J.R. (1994). Sampling theory for Paley–Wiener spaces in the Riesz basis setting. *Proc. R. Irish Acad.*, **94A**, 219–236.

63. Higgins, J.R. Sampling for multi-band functions. (1994). In: *Proceedings of the International conference on Mathematical Analysis and Signal Processing*. Cairo, Jan. 1994, pp 165–170. Contemporary Mathematics Series, AMS.

64. Hille, E. (1959). *Analytic function theory I*. Ginn and Company, Boston.

65. Hinsen, G. (1992). Explicit irregular sampling formulas. *Comput. Math. Appl.*, **40**, 177–198.

66. Hinsen, G. (1993). Irregular sampling of bandlimited L^p functions. *J. Approx. Theory*, **72**, 346–364.

67. Hinsen, G. and Klösters, D. (1993). The sampling series as a limiting case of Lagrange interpolation. *Appl. Analysis*, **49**, 49–60.

68. Horgan, J. (1992). Claude E. Shannon. *IEEE Spectrum*, **29**, 72–75.

69. Hörmander, L. (1964). *Linear partial differential operators*. Springer-Verlag, Berlin.

70. Hoskins, R.F. and Sousa Pinto, J. (1994) *Distributions, ultradistributions and other generalised functions*. Ellis Horwood, Chichester.

71. Hruščev, S.V., Nikol'skii, N.K. and Pavlov, B.S. (1981). Unconditional bases of exponentials and of reproducing kernels. In: *Complex analysis and spectral theory*, Seminar, Leningrad 1979/80 (eds V.P. Havin and N.K. Nikol'skii). Lecture Notes in Mathematics **864**. Springer-Verlag, Berlin, 214–335.

72. Jaffard, S. (1991). A density criterion for frames of complex exponentials. *Mich. Math. J.*, **38**, 339–348.

73. Jerri, A. J. (1977). The Shannon sampling theorem—its various extensions and applications: a tutorial review. *Proc. IEEE*, **65**, 1565–1596.

74. Jerri, A.J. (1992). *Integral and discrete transforms with applications and error analysis*. Marcel Dekker, New York.

75. Jerri, A.J. (1993). Error analysis in application of generalizations of the sampling theorem. In: *Advanced topics in Shannon sampling and interpolation theory* (ed. R.J. Marks II). Springer-Verlag, New York, 218–285.

76. Kempski, B.L. (1995). Extensions of the Whittaker–Shannon sampling series aided by symbolic computation. M. Phil. Thesis, Anglia Polytechnic University, Cambridge.

77. Kluvánek, I. (1965). Sampling theorem in abstract harmonic analysis. *Mat.-Fyz. Časopis Sloven. Akad. Vied*, **15**, 43–48.

78. Kohlenberg, A. (1953). Exact interpolation of band-limited functions. *J. Appl. Phys.*, **24**, 1432–1436.

79. Kotel'nikov, V.A. (1933). On the carrying capacity of the "ether" and wire in telecommunications. Material for the first all-union conference on questions of communication (Russian). Izd. Red. Upr. Svyazi RKKA, Moscow.

80. Kramer, H. P. (1957). A generalised sampling theorem. *J. Math. Phys.*, **63**, 68–72.

81. Landau, H.J. (1967a). Sampling, data transmission, and the Nyquist rate. *Proc. IEEE*, **55**, 1701–1706.

82. Landau, H.J. (1967b). Necessary density conditions for sampling and interpolation of certain entire functions. *Acta Math.*, **117**, 37–52.

83. Landau, H.J. (1985). An overview of time and frequency limiting. In: *Fourier techniques and applications* (ed. J.F. Price). Plenum Press, New York.

84. Loomis, L.H. (1953). *An introduction to abstract harmonic analysis*. Van Nostrand, New York.

85. Lorch, E.R. (1962). *Spectral theory*. Oxford University Press, New York.

86. Lüke, H.D. (1977). Zeitinvariante abtast- und PAM-systeme. *Archiv Elektr. Übertragungstechnik*, **31**, 441–445.

87. Lüke, H.D. (1978). Zur entstehung des Abtasttheorems. *Nachr. Tech. Z.*, **31**, 271–274.

88. Lyubarskiĭ, Y.I. and Seip, K. (1994). Sampling and interpolation of entire functions and exponential systems in convex domains. *Ark. Mat.*, **32**, 157–193.

89. Magnus, W., Oberhettinger, F. and Soni, R.P. (1966). *Formulas and theorems for the special functions of mathematical physics*. Springer-Verlag, Berlin.

90. Marks II, R.J. (1991). *Introduction to Shannon sampling and interpolation theory*. Springer-Verlag, New York.

91. Marks II, R.J. (ed.) (1993). *Advanced topics in Shannon sampling and interpolation theory*. Springer-Verlag, New York.

92. Marti, J.T. (1969). *Introduction to the theory of bases*. Springer Tracts in Natural Philosophy, Vol. 18. Springer-Verlag, New York.

93. Marvasti, F. (1993). Nonuniform sampling. In: *Advanced topics in Shannon sampling and interpolation theory* (ed. R.J. Marks II). Springer-Verlag, New York, 121–156.

94. Meyer, Y. (1993). *An introduction to wavelets* by Charles K. Chui; *Ten lectures on wavelets* by Ingrid Daubechies — reviewed by Yves Meyer. *Bull. Amer. Math. Soc.*, **28**, 350–360.

95. Mihailov, V.P. (1962). Riesz bases in $L_2(0,1)$, *Sov. Math.* (English translation: *Dokl., Acad. Sci. U.S.S.R.*) **3**(3), 851–855.

96. Mills, S. (1985). The independent derivations by Leonhard Euler and Colin MacLaurin of the Euler–MacLaurin summation formula. *Arch. Hist. Exact*

Sci., **33**(1), 1–13.

97. Mugler, D.H. (1976). Convolution, differential equations, and entire functions of exponential type. *Trans. Amer. Math. Soc.*, **216**, 145–187.

98. Nashed, M.Z. and Walter, G.G. (1991). General sampling theorems for functions in reproducing kernel Hilbert spaces. *Math. Control Signals Systems*, **4**, 363–390.

99. Nielsen, N. (1904). *Handbuch der Cylinderfunktionen.* Teubner, Leipzig.

100. Nikol'skiĭ, S.M., (1975). *Approximation of functions of several variables and imbedding theorems.* Springer-Verlag, Berlin.

101. Oberhettinger, F. (1972).   *Tables of Bessel transforms.* Springer-Verlag, Berlin.

102. Paley, R.E.A.C. and Wiener, N. (1934). *Fourier transforms in the complex domain.* Am. Math. Soc. Colloq. Publ. 19.

103. Papoulis, A. (1977). *Signal analysis.* McGraw-Hill, New York.

104. Pichler, F. (1973). Walsh functions—introduction to the theory. In: *Signal Processing (Proc. NATO Advanced Study Institute on Signal Processing)* (eds J.W.R. Griffiths *et al.* Academic Press, New York.

105. Pollak, H.O. (1963). Panel No. 3, New directions in applied mathematics. In: *Dartmouth College Mathematics Conference, 1961.* New directions in mathematics (ed. R.W. Ritchie), Prentice-Hall, Englewood Cliffs, NJ, 68–77.

106. Raabe, H. (1939).   Untersuchungen an der wechselseitigen Mehrfachübertragung (Multiplexübertragung). *Elek. Nachrichtentechnik*, **6**, 213–228.

107. Rahman, Q.I. and Schmeißer, G. (1985). Reconstruction and approximation of functions from samples. In *Delay equations, approximation and aplication* (eds G. Meinardus and G. Nürnberger). International Series of Numerical Mathematics, Vol. 74, Birkhäuser-Verlag, Basel, 213–233.

108. Rahman, Q.I. and Schmeißer, G. (1994). The summation formulae of Poisson, Plana, Euler–Maclaurin and their relationship. *J. Math. Sci. (Part I)*, **28**, 151–171.

109. Rawn, M.D. (1989). A stable nonuniform sampling expansion involving derivatives. *IEEE Trans. Inform. Theory*, **35**, 1223–1227.

110. Riesz, F. and Sz.-Nagy, B. (1990). *Functional analysis* (2nd edn) (tr. L.F. Boron). Dover, New York.

111. Rochberg, R. (1987). Toeplitz and Hankel operators on the Paley-Wiener space. *Integral Equations Operator Theory*, **10**, 187–235.

112. Saitoh, S. (1988). *Theory of reproducing kernels and its applications.* Longman, Harlow.

113. Sansone, G. (1959). *Orthogonal functions.* Interscience, New York.

114. Seip, K. (1987). An irregular sampling theorem for functions bandlimited in a generalized sense. *SIAM J. Appl. Math.*, **47**, 1112–1116.

115. Seip, K. (1992). Density theorems for sampling and interpolation in the Bargmann–Fock space. *Bull. Am. Math. Soc.*, **26**, 322–328.

116. Seip, K. (1995a). A simple construction of exponential bases in L^2 of the union of several intervals. *Proc. Edinb. Math. Soc.*, **38**, 171–177.

117. Seip, K. (1995b). On the connection between exponential bases and certain related sequences in $L^2(-\pi, \pi)$. *J. Functional Analysis*, **130**, 131–160.

118. Shannon, C.E. (1949). Communication in the presence of noise. *Proc. IRE*, **37**, 10–21.

119. Singer, I. (1970). *Bases in Banach spaces I*. Springer-Verlag, Berlin.

120. Slepian, D. (1976). On bandwidth. *Second Shannon Lecture. Proc. IEEE*, **64**, 292–300.

121. Slepian, D. (1983). Some comments on Fourier analysis, uncertainty and modeling. *SIAM Rev.*, **25**, 379–393.

122. Slepian, D. and Pollak, H.O. (1961). Prolate spheroidal wave functions, Fourier analysis and uncertainty, I. *Bell Systems Tech. J.*, **40**, 43–64.

123. Smith, K.T., Solomon, D. C. and Wagner, S. L. (1977). Practical and mathematical aspects of the problem of reconstructing objects from radiographs. *Bull. Am. Math. Soc.*, **83**, 1227–1270.

124. Splettstößer, W. (1978a). Some extensions of the sampling theorem. In: *Linear Spaces and Approximation (Proc. Conf. Res. Inst. Oberwolfach, 1977)* (eds P.L. Butzer and B. Sz.-Nagy). Birkhäuser-Verlag, Basel, 615–628.

125. Splettstößer, W. (1978b). On generalized sampling sums based on convolution integrals. *Archiv für Elektronik und Übertragungstechnik*, **32**, 267–275.

126. Splettstößer, W. (1983). 75 years alising error in the sampling theorem. In: *EUSIPCO - 83, Signal processing: theories and applications (Second European signal processing Conference, Erlangen, 1983)* (ed. H.W. Schüssler). North-Holland, Amsterdam, 1–4.

127. Splettstößer, W., Stens, R.L. and Wilmes, G. (1981). On approximation by the interpolating series of G. Valiron. *Funct. Approximatio Comment. Math.*, **11**, 40–56.

128. Standish, C.J. (1967). Two remarks on the reconstruction of sampled non-bandlimited functions. *IBM J. Res. Develop.*, **11**, 648–649.

129. Stanković, R.S., Stojić, M.R., and Bogdanović, S.M. (1988). *Fourierovo predstavljanje signala*. Naučna Knjiga, Beograd.

130. Stein, E.M. (1976). Harmonic analysis on R^N. In: *Studies in harmonic analysis* (ed. J. M. Ash), Studies in Mathematics, Vol. 13. Mathematical Association of America, Washington DC, 97–135.

131. Stein, E.M. and Weiss, G. (1971). *Introduction to Fourier analysis on Euclidean spaces*. Princeton University Press, Princeton, New Jersey.

132. Stenger, F. (1981). Numerical methods based on Whittaker cardinal, or sinc functions. *SIAM Rev.*, **23**, 165–224.

133. Stenger, F. (1993). *Numerical methods based on sinc and analytic functions*. Springer Series in Computational Mathematics 20. Springer-Verlag, Berlin.

134. Stens, R.L. (1983). A unified approach to sampling theorems for derivatives and Hilbert transforms. *Signal Proc.*, **5**, 139–151.

135. Stickler, D.C. (1967). An upper bound on aliasing error. *Proc. IEEE*, **55**, 418–419.

136. Strang, G. (1993). Wavelet transforms versus Fourier transforms. *Bull. Am. Math. Soc.*, **28**, 288–305.

137. Titchmarsh, E.C. (1948). *Introduction to the theory of Fourier integrals* (2nd edn). Clarendon Press, Oxford.

138. Titchmarsh, E.C. (1962). *Eigenfunction expansions associated with second-order differential equations part 1* (2nd edn). Clarendon Press, Oxford.

139. Vaaler, J.D. (1985). Some extremal functions in Fourier analysis. *Bull. Am. Math. Soc.*, **12**, 183–216.

140. Walker, W.J. (1991a). Zeros of the Fourier transform of a distribution. *J. Math. Analysis Applics*, **154**, 77–79.

141. Walker, W.J. (1991b). Almost periodic functions with bounded Fourier exponents. *J. Math. Anal. Appl.*, **162**, 537–541.

142. Walker, W.J. (1994). Oscillatory properties of Paley–Wiener functions. *Ind. J. Pure Appl. Math.*, **25**, 1253–1258.

143. Walter, G.G. (1994). *Wavelets and other orthogonal systems with applications*. CRC Press, Boca Raton, FL.

144. Waring, E. (1779). Problems concerning interpolations. *Phil. Trans. R. Soc.*, **69**, 59–67.

145. Weston, J.D. (1949). The cardinal series in Hilbert space. *Proc. Camb. Phil. Soc.*, **45**, 449–453.

146. Whiteside, D. T. (1961). Patterns of mathematical thought in the later seventeenth century. *Arch. Hist. Exact Sci.*, **1**, 178–388.

147. Whittaker, E. T. (1915). On the functions which are represented by the expansions of the interpolation theory. *Proc. R. Soc. Edinb.*, **35**, 181–194.

148. Whittaker, E.T. and Watson, G.N. (1962). *A course of modern analysis* (4th edn). Cambridge University Press.

149. Whittaker, J.M. (1935). *Interpolatory function theory.* Cambridge University Press.

150. Woodward, P.M. (1953). *Probability and information theory, with applications to radar*. Pergamon Press, Oxford.

151. Yao, K. (1967). Applications of reproducing kernel Hilbert spaces—band-limited signal models. *Information and Control*, **11**, 429–444.

152. Young, R.M. (1980). *An introduction to nonharmonic Fourier series.* Academic Press, New York.

153. Zayed, A.I. (1991). On Kramer's sampling theorem associated with general Sturm–Liouville problems and Lagrange interpolation. *SIAM J. Appl. Math.*, **51**, 575–604.

154. Zayed, A. I. (1993a). A proof of new summation formulae by using sampling theorems. *Proc. Am. Math. Soc.*, **117**, 699–710.

155. Zayed, A. I. (1993b). *Advances in Shannon's sampling theory.* CRC Press, Boca Raton, FL.

156. Zayed, A.I., Hinsen, G. and Butzer, P.L. (1990). On Lagrange interpolation and Kramer-type sampling theorems associated with Sturm–Liouville problems, *SIAM J. Appl. Math.* **50**, 893–909.

157. Zwaan, M. (1990a). Error estimates for nonuniform sampling. *Num. Funct. Analysis Optim.*, **11**(5 and 6), 589–599.

158. Zwaan, M. (1990b). Approximation of the solution to the moment problem in a Hilbert space. *Num. Funct. Analysis Optim.*, **11**(5 and 6), 601–608.

159. Zwaan, M. (1995). Bounds for errors in nonuniform sinc interpolation. In: *1995 workshop on sampling theory and applications.* Conference Proceedings, Jurmala, Latvia, Sept. 1995, pp 55–58. Institute of Electronics and Computer Science, Riga, Latvia.

INDEX